IN THE SHADOWS OF THE TROPICS

T0248602

Re-materialising Cultural Geography

Dr Mark Boyle, Department of Geography, University of Strathclyde, UK and Professor Donald Mitchell, Maxwell School, Syracuse University, USA

Nearly 25 years have elapsed since Peter Jackson's seminal call to integrate cultural geography back into the heart of social geography. During this time, a wealth of research has been published which has improved our understanding of how culture both plays a part in, and in turn, is shaped by social relations based on class, gender, race, ethnicity, nationality, disability, age, sexuality and so on. In spite of the achievements of this mountain of scholarship, the task of grounding culture in its proper social contexts remains in its infancy. This series therefore seeks to promote the continued significance of exploring the dialectical relations which exist between culture, social relations and space and place. Its overall aim is to make a contribution to the consolidation, development and promotion of the ongoing project of re-materialising cultural geography.

Other titles in the series

Geographies of Muslim Identities
Diaspora, Gender and Belonging
Edited by Cara Aitchison, Peter Hopkins and Mei-Po Kwan
ISBN 978 0 7546 4888 8

Presenting America's World
Strategies of Innocence in National Geographic Magazine, 1888–1945
Tamar Y. Rothenberg
ISBN 978 0 7546 4510 8

Regulating the Night
Race, Culture and Exclusion in the Making of the Night-time Economy
Deborah Talbot
ISBN 978 0 7546 4752 2

Mixed Towns, Trapped Communities
Historical Narratives, Spatial Dynamics, Gender Relations
and Cultural Encounters in Palestinian-Israeli Towns
Edited by Daniel Monterescu and Dan Rabinowitz
ISBN 978 0 7546 4732 4

In the Shadows of the Tropics
Climate, Race and Biopower in Nineteenth Century Ceylon

JAMES S. DUNCAN
Emmanuel College, University of Cambridge, UK

Routledge
Taylor & Francis Group

LONDON AND NEW YORK

First published 2007 by Ashgate Publishing

2 Park Square, Milton Park, Abingdon, Oxon OX14 4RN
711 Third Avenue, New York, NY 10017, USA

Routledge is an imprint of the Taylor & Francis Group, an informa business

First issued in paperback 2016

British Library Cataloguing in Publication Data
Duncan, James S.
  In the shadows of the tropics : climate, race and biopower
  in nineteenth century Ceylon. - (Re-materialising cultural
  geography)
  1. Coffee plantations - Sri Lanka - History - 19th century
  2. British - Sri Lanka - Social conditions - 19th century
  3. Plantation life - Sri Lanka - History - 19th century
  4. Sri Lanka - Social conditions - 19th century 5. Sri
  Lanka - Ethnic relations - History - 19th century
  I. Title
  338.1'7373'095493'09034

Library of Congress Cataloging-in-Publication Data
Duncan, James S.
  In the shadows of the tropics : climate, race and biopower in nineteenth century
Ceylon / by James S. Duncan.
      p. cm. (Re-materialising cultural geography)
  Includes bibliographical references and index.
  ISBN 978-0-7546-7226-5
  1. Coffee industry--Sri Lanka--History--19th century. 2. Great Britain--Colonies--
History. 3. Race. I. Title.

  HD9199.S722D86 2007
  338.1'737309549309034--dc22

                                                                  2007013158

ISBN 13: 978-0-7546-7226-5 (hbk)
ISBN 13: 978-1-138-26241-6 (pbk)

Contents

For Nancy

List of Maps and Figures

List of Tables

Acknowledgements

I am indebted to Gerald Peiris for inviting me to be a visiting research fellow in the Department of Geography, University of Peradeniya. I also owe a great debt of gratitude to Nalini and Shantha Hennayake for the intellectual stimulation and friendship they provided both in Sri Lanka and North America. I am appreciative of the help rendered by the staffs at the Ceylon Room, University of Peradeniya Library, the Kandy City Archives, and the Sri Lanka National Archives for their helpfulness in locating material. I would like to thank Terri Barringer, whose knowledge of the Royal Commonwealth Society Collection is unmatched and to Jo Doake, who for a time served as my research assistant. I am also grateful to the staffs of the Centre for South Asia, University of Cambridge, Official Publications Room and Rare Books Room of Cambridge University Library, the National Archives at Kew (formerly the Public Record Office) and The National Library of Scotland, Department of Manuscripts. I am indebted also to Philip Stickler and Owen Tucker of the Department of Geography, University of Cambridge for producing the maps.

Over the years I have presented preliminary versions of the material contained here in seminars and at conferences in Europe and North America. I am grateful for the feedback that I have received. In particular I would like to thank the following colleagues and friends for their intellectual stimulation and support: John Agnew, Trevor Barnes, Rana Behal, Peter Burke, Stuart Corbridge, Liz Gagen, Phil Howell, Tariq Jazeel, Nuala Johnson, Gerry Kearns, Judith Kenny, Satish Kumar, Dave Lambert, Michael Landzelius, David Ley, David Robinson, Mitch Rose, Rich Schein, Heidi Scott, Jo Sharp, Kapila Silva, Richard Smith, and Sujit Sivasundaram. I owe particular thanks to Steve Legg, for it was through the influence of his work on colonial Delhi that I developed an interest in the study of colonial governmentality. I am grateful to Maria-Lucia Pallares-Burke and Peter Burke, John and Julietta Harvey, Larry Klein and the whole De Menthon family for their friendship in Cambridge. My greatest debt is to Nancy Duncan who has discussed all of the ideas in this book with me and whose editorial help as ever has proved invaluable.

An earlier version of Chapter 3 appeared in J.S. Duncan, 'The Struggle to be Temperate: Climate and "Moral Masculinity" in Mid-Nineteenth Century Ceylon', *Singapore Journal of Tropical Geography* 21 (2000): 34–47; J.S. Duncan, 'Sombres Pensées Dans la Maison Coloniale: Masculinité, Contrôle et Refoulement Domestiques à Ceylan au Milieu du XIXème Siècle', in *Espaces Domestiques. Construire, Aménager, Représenter*, edited by B. Collignon and J.-F. Staszak (Paris: Bréal, 2004), 341–53; and J.S. Duncan, 'Home alone? Masculinity, Discipline and Erasure in Mid-Nineteenth Century Ceylon', in *Gendered Landscapes*, edited by L. Dowler et al. (New York: Routledge, 2005), 19–33. An earlier version of Chapter 4 appeared in J. Duncan, 'Embodying Colonialism?: Domination and Resistance in

19th Century Ceylonese Coffee Plantations', *Journal of Historical Geography* 28 (2002): 317–38.

Finally, I am indebted to my commissioning editor Val Rose and my desk editor, Pam Bertram for seeing this manuscript so painlessly through the production process.

List of Abbreviations

AGA   Assistant Government Agent
ARC   Administration Reports, Ceylon
BPP   British Parliamentary Papers
CH   Ceylon Hansard [Debates of the Ceylon Legislative Council]
CO   Colonial Office
GA   Government Agent
PCAS   Proceedings of the Ceylon Agricultural Society
PPA   Proceedings of the Planters' Association
PRO   Public Record Office (Kew)
SP   Sessional Papers

Glossary of Terms

Aratchi: A village headman.

Assistant Government Agent (AGA): Official in charge of a district.

Barbeque: The floor where coffee beans are dried.

Boutique: A roadside store.

Chena: Shifting cultivation of forest land.

Chetty (Chettiar): An Indian money lender and businessman.

Coffee Gardens: Small holder coffee plots, as distinguished from Estates.

Coir: The fibre from the husk of the coconut used to make mats, twine, etc. From the Tamil word *kayaru* (cord).

Cooly: A labourer on an estate or in the Public Works Department. They were normally but not invariably Tamil immigrants.

Country-bottled: A derogatory term for an English person who was born in the tropics. More polite variants were country-born, and country-bred.

Estates: The term used in Ceylon for coffee plantations.

Government Agent (GA): Official in charge of a province.

Kanakapillai: Estate accountant.

Kandyan: Those Sinhalese people living in the highland regions of what constituted the Kandyan Kingdom.

Kangany: A Tamil supervisor of a labour gang.

Kurakkan: Millet (*Eleusine coracana*).

Malabar: The name given locally to Tamil migrant labourers.

Moor: The names given to Muslims living in Ceylon.

Mudaliyar: Chief headman.

Oakum picking: Separating threads from discarded naval rope. The threads were used to make string or stuff mattresses.

Paddy: Unhusked rice.

Parchment Coffee: Coffee beans from which a fine membrane has yet to be removed.

Patana: Grasslands

Peon: A labourer.

Periya dorai: 'Big Master', Tamil name for a planter.

Pulping House: The place where coffee beans are separated from the fruit (cherries).

Rajakariya: Compulsory service due to the state.

Sesilara kangany: Assistant kangany.

Sinhalese: One of the major ethnic groups of Ceylon.

Sinna Dorai: 'Little Master', Tamil name for the Assistant Planter.

Tamil: A native of south India.

Tavalam: A pack animal (usually cattle).

Vidahn: Local headman with responsibility for law and order.

Chapter 1

Introduction

A fear haunted ... [those who ruled]: the fear of darkened spaces, of the pall of gloom which prevents the full visibility of things, men and truths. It sought to break up the patches of darkness that blocked the light, eliminate the shadowy areas of society, demolish the unlit chambers where arbitrary political acts, monarchical caprice, religious superstition, tyrannical and priestly plots, epidemics and the illusions of ignorance were fomented.

Michel Foucault[1]

The colonies ... were underfunded and overextended laboratories of modernity. There, science's authority as a sign of modernity was instituted with a minimum of expense and a maximum of ambition.

Gyan Prakash[2]

Introduction

The tropics cast long shadows over the practices of colonialism in Ceylon during the nineteenth century. Shadows were cast by the very materiality of the tropical environment itself; the heat and torrential rains, the dark, dank forests, and the mysterious diseases of humans, animals and plants. And shadows were cast also by the European imagination of tropical environments and peoples. Although never thought of as a true 'heart of darkness', a description reserved for West Africa, Ceylon was nevertheless viewed as only very partially visible and hence only partially controllable by Europeans. While darkness was a classic trope of the broader colonising mission with its attendant images of European modernity bringing enlightenment to occluded corners of the world, the metaphor of shadows suggests the ambiguity of a region where a flickering light is seen to penetrate only partially or sporadically.[3] The metaphor of light and shadow thus captures my intended focus on the difficulties of setting up modern forms of bureaucracy, governmentality and biopower which depended conceptually upon the hegemony of vision and knowledge

---

1 M. Foucault, 'The Eye of Power', in *Power/Knowledge: Selected Interviews and Other Writings, 1972–1977*, edited by C. Gordon (New York: Pantheon, 1980), 153.

2 G. Prakash, *Another Reason: Science and the Imagination of Modern India* (Princeton: Princeton University Press, 1999), 13.

3 The history of light as a metaphor for certainty, legibility and mastery, both technical and moral, is well documented in D. Levin, (ed.) *Modernity and the Hegemony of Vision* (Berkeley: University of California Press, 1993). See S. Houlgate, 'Vision, Reflection, and Openness', in *Modernity and the Hegemony of Vision*, edited by D. Levin, 87–123. (Berkeley: University of California Press, 1993) for a discussion of shadow as a more supple and subtle metaphor for shortsightedness, and elusive or ambiguous knowledge.

necessary for political domination and mastery over nature.[4] The idea of shadows also conveys the ambiguity and tensions between racialized and biologized[5] social hierarchies and liberal democratic modernity. While modernity was in the throes of dismantling naturalized racial differences and grappling with evolving ideas of citizenship and equality, the tropics cast shadows of doubt over these universalizing ideals. Widespread racial prejudices were supported by what the British took as empirical evidence of racial and environmentally-determined difference. The racism that coloured their experience in the tropics was further supported by theories of environmental determinism which remained the scientific orthodoxy of the day. In varied ways the mysterious material disease ecologies and their own colonial mentalities worked to obscure the tropics and tropical societies and practices for the British. The fear of racial degeneration hung darkly over their lives in Ceylon. Shadows were also cast deliberately by some of the colonists themselves as they resisted attempts by the state to make their own, often ruthless, deployment of power and management of the lives and deaths of labouring bodies more legible and accountable. By focusing upon shadows rather than light, I wish to draw attention to the contested, fragmentary and often ineffective nature of colonial practices, in particular the practices of labour reproduction on the British-owned coffee estates. I also want to explore the many 'plans gone wrong' that brought unintended misery and often devastation to planters, plantation labourers and peasants alike.

Nicolas Thomas calls upon historians of colonialism to be more attentive to failure. In his own research he uses the failures of colonists in the Pacific as a focal point for exploring the complexity of the vast set of loosely interconnected projects we call colonialism, and to deflate the ideology of colonial power as a totalizing and crushing structure.[6] While Scott shows the diffuseness of power by exploring the 'weapons of the weak,'[7] Thomas looks at failure among those conventionally associated with colonial power. He writes ethnographies of 'bad colonists,' exploring the 'non-correspondence' between the ideology of colonialism and the self-accounts of those who failed in their colonial endeavours. I write not so much about personal failures as about systemic failure within the realms of agriculture, medicine, crime prevention, and political legitimacy more generally. Ultimately, the book is about

---

4   For reviews of work in geography on governmentality see R. Rose-Redwood, 'Governmentality, Geography and the Geo-Coded World', *Progress in Human Geography* 30 (2006): 469–86. See also J. Crampton and S. Elden (eds.) *Space, Knowledge, and Power: Foucault and Geography* (Aldershot, Ashgate, 2007) for a range of effective papers using the idea of governmentality.

5   See N. Rose and C. Novas 'Biological Citizenship', in *Global Assemblages: Technology, Politics, and Ethics as Anthropological Problems*, edited by. A. Ong and S. Collier, 439–463. (Oxford: Blackwell, 2005) on the biologization of the concept of citizenship and human rights that during the nineteenth century included beliefs and disputes about racialized moral capacities, racial deterioration and degeneracy.

6   N. Thomas, *Colonialism's Culture: Anthropology, Travel, and Government* (Cambridge: Polity, 1994), 166–67; N. Thomas and R. Eves, *Bad Colonists: The South Seas Letters of Vernon Lee Walker and Louis Becke* (Durham: Duke University Press, 1999).

7   Scott, J. *Weapons of the Weak: Everyday Forms of Peasant Resistance* (New Haven: Yale University Press, 1985).

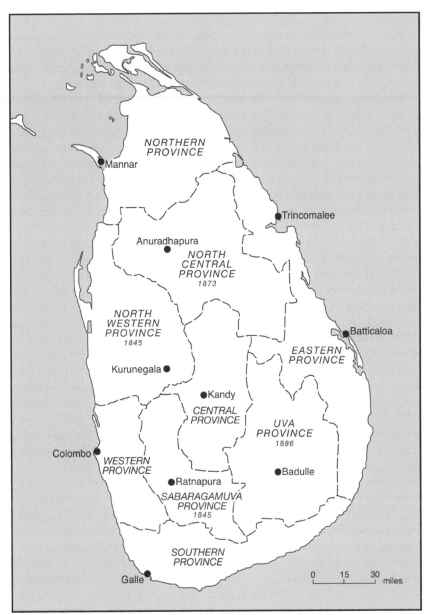

Map 1.1   Ceylon provinces in 1889

Source: Adapted from I.H. Vanden Driesen, *Indian Plantation Labour in Sri Lanka: Aspects of the History of Immigration in the 19th Century* (Nedlands: University of Western Australia, 1982).

the inability of agents to concentrate power. Sites of power lay dispersed throughout the colony and the global assemblages into which they were integrated. Varying degrees of and types of power were wielded not only by estate owners and colonial government officials, but also by the estate labourers who foot-dragged, the villagers who crept onto the estates at night, the tropical climate, and a host of diseases such as the cholera bacilli, and the *Hemileia vastatrix*, the coffee disease that destroyed the livelihood of tens of thousands of people and continues relatively unchecked around the world 140 years after it was first identified in the Kandyan highlands.

There were many forms of resistance to concentrations of colonial power on the part of human and non-human agents alike. Planters and those whose interests lay with the success of the coffee estates in the highlands were very aware of the ability of the non-human, be it wind, rain, drought, or disease to frustrate their plans. In fact, under the influence of environmental determinism and theories of tropical degeneration, mid-nineteenth century European imaginations ceded to the physical environment even more power than it in fact had. Additionally, as I will demonstrate throughout this book, Europeans encountered continuous, usually covert human resistance to their projects. And often, resistance which does not announce itself is the most effective. As Gordon puts it, 'the existence of those who seem not to rebel is a warren of minute, individual, autonomous tactics and strategies which counter and inflect the visible facts of overall domination...'.[8] Accordingly, this book is a study of the diffuseness of power and countervailing powers within a broad framework of domination. British colonialism and the colonial plantation economy are presented as frustrated dreams of ordering tropical nature and non-European lives and labour.

Doreen Massey calls for the spatialization of the history of modernity away from 'the unfolding, internal story of Europe alone.'[9] She sees modernity as composed of many coeval, intersecting trajectories and networks. Modernity was always necessarily locally differentiated, fractured and negotiated. It makes little sense to speak of modernity as a unified project or even a discourse or regime of practices. Modernity and its various manifestations such as capitalism and governmentality were always multiple and spatially differentiated. Knowledge practices and techniques flowed into and through the colonies and were transformed there not only as a new imposed order, but as newly evolving local and extra-local forms. My wish then is to contribute to research that, as Timothy Mitchell puts it, theorizes the question of modernity in a way that 'relocates it within a global context and, at the same time enables that context to complicate, rather than simply reverse the narrative logic of modernization.'[10]

Along with authors such as David Scott, Ann Stoler, and Talal Asad, who are interested in revealing the spatial unevenness and fragmentation of modernity through a focus on colonial governmentality and biopower, I see the practices of colonial power, and the difference that race and tropical climates were thought to make as central to the working out of governmental rationalities. The production of colonial knowledge with its focus on the racialized working body and the material conditions in which that body was to live or die was central to developing notions

---

8   Colin Gordon quoted in J. Scott, *Weapons of the Weak*, frontispiece.

9   D. Massey, *For Space* (London: Sage, 2005), 63.

10 T. Mitchell, 'The Stage of Modernity', in *Questions of Modernity*, edited by T. Mitchell (London: University of Minnesota Press), 2000, 7.

of governmentality.[11] In line with the broad project of opening out the history of modernity spatially and repositioning it within intersecting global trajectories, I hope that this book will contribute to the larger story of modern governmentality some of the complexity and differentiation that a focus on Europe alone obscures. The idea of trajectories and networks, according to David Lambert, 'allows us to treat the metropole and colonies as interconnected analytical fields – which is emphatically not to say that the two are "the same" but that each provides a local translation of a wider imperial circuit that impacted the forms of labour, consumption, servitude, freedom and belonging in specific ways.'[12]

Prakash has introduced the question of whether colonialism was the 'tropicalization' of modernity.[13] I see tropicalization as an absolutely key term in the debate that Prakash addresses. The British who colonized Ceylon took the idea of tropicality and tropical difference very seriously.[14] How were modern, liberal ideals to be translated into a colonial context where the colonized had no modern citizenship rights? The question of what race is and what difference it makes to 'natural rights' was not seen in the nineteenth century as a question of biology, or at least not of biology alone. Nor was it primarily about culture in the sense of customs and mores. Rather, it was seen as fundamentally about environment – and in the case of Ceylon – the tropical environment.

Modernity in the tropics was ambiguous to the British. As Peter Redfield puts it, the colonial vision of modernity 'includes not only the greater apparatus of normalization but also the shadow of its partial exceptions.'[15] Thus, this book is about how the British were thinking about the tropics and tropical difference – environmental, racial, cultural, moral difference. It is about the shadows cast over modernity by the (partial) exceptions that followed from differences, as they saw them. Work by

---

11 See T. Asad, 'Conscripts of Western Civilization', in *Dialectical Anthropology: Essays in Honor of Stanley Diamond, Vol. 1 Civilization in Crisis*, edited by C. Gailey (Gainsville: University Press of Florida, 1992), 337. D. Scott, 'Colonial Governmentality', in *Anthropologies of Modernity: Foucault, Governmentality and Life Politics*, edited by J. Inda, 23–49. (Oxford: Blackwell, 2005), 30; A.L. Stoler, *Carnal Knowledge and Imperial Power: Race and the Intimate in Colonial Rule* (Berkeley: University of California Press, 2002).

12 K. Wilson, *The Island Race: Englishness, Empire and Gender in the Eighteenth Century* (London: Routledge, 2003). Quoted in D. Lambert, *White Creole Culture, Politics and Identity During the Age of Abolition* (Cambridge: Cambridge University Press, 2005), 22. This point is also effectively made by A. Lester, 'Constituting Colonial Discourse', in *Post-Colonial Geographies*, edited by A. Blunt and C. McEwan, 29–45. (New York: Continuum, 2002).

13 Prakash, *Another Reason*, 125–6.

14 On tropicality see F. Driver and B. Yeoh, 'Constructing the Tropics: Introduction', *Singapore Journal of Tropical Geography* 21 (2000): 1–5; F. Driver, 'Imagining the Tropics: Views and Visions of the Tropical World', *Singapore Journal of Tropical Geography* 25 (2004): 1–17; G. Bowd and D. Clayton, 'Fieldwork and Tropicality in French Indochina: Reflections on Pierre Gourou's "Les paysans du Delta Tonkinois, 1937"', *Singapore Journal of Tropical Geography* 41 (2003): 147–68.

15 P. Redfield, 'Foucault in the Tropics: Displacing the Panopticon', in *Anthropologies of Modernity: Foucault, Governmentality and Life Politics*, edited by J.X. Inda. (Oxford: Blackwell, 2005), 67.

geographers such as David Nally and Stephen Legg[16] on colonial governmentality has begun to explore race as an important basis of internal differentiation within populations to be governed. Foucault believed that racism tended to 'fragment, to create caesuras within the biological continuum addressed by biopower.'[17] This book is not only about the racialization of populations, it is also about the how race was understood within the context of nineteenth century environmental determinism and theories of tropical degeneration. I have followed the story of those who were actually working out theories of colonial governmentality 'on the ground' and I have found that they viewed the interlinked issues of climate and race as absolutely central to their concerns. This materiality of nature was central to what Clayton[18] has called, the 'messy pragmatics of colonial contact.'

This book traces the rise and fall of coffee production in highland Ceylon from the 1830s to the 1880s, showing at least in broad outline how the plantation system was constituted by the meeting up of histories through interpenetrating networks of nature/science/governmentality/culture. I hope to show through the story of coffee that there are many different human agents, discourses, technologies, and rationalities of government that co-exist with non-human agents, material preconditions, spatialities, and particular historical trajectories that must be considered in order to undermine the story of capitalist modernity as centred in Europe. Coffee, which had been an important world commodity since the sixteenth century, had gathered around itself a vast assemblage of sites, routes, ecologies, technologies, and human and non-human agents. It was a good produced in the tropics largely for consumption in the temperate zone. Although until the mid-nineteenth century it was primarily a colonial crop, the fact that it later became associated with independent Latin American countries has resulted in colonial coffee production remaining understudied.[19] Coffee estates in the Ceylonese highlands were particularly interesting sites of experiments in modern bio-political practices of power during the nineteenth century.

As much of the literature on governmentality has been urban in focus, there has been relatively less attention given to early efforts to expand governmentality into rural economies. When work on governmentality in geography and related fields

---

16 S. Legg, *Spaces of Colonialism: Discipline and Governmentality in Delhi, India's New Capital* (Oxford: Blackwell, 2007). S. Legg, 'Governmentality, Congestion and Calculation in Colonial Delhi', *Social and Cultural Geography* 7 (2006): 709–29. S. Legg, 'Foucault's Population Geography: Classifications, Biopolitical and Governmental Spaces', *Population, Space and Place* 2 (2005): 137–56. D. Nally, '"Eternity's Commissioner"; Thomas Carlyle, the Great Irish Famine and the Geopolitics of Travel', *Journal of Historical Geography* 32 (2006): 313–335.

17 M. Foucault, *The Essential Foucault*, edited by N. Rose. (New York: New Press, 2003), 255.

18 D. Clayton, 'Imperial Geographies', in *A Companion to Cultural Geography*, edited by J.S. Duncan, N.C. Johnson and R.H. Schein. (Oxford: Blackwell, 2004), 458. Clayton's *Islands of Truth: The Imperial Fashioning of Vancouver Island* (Vancouver: University of British Columbia Press, 2000) is a superb study of the pragmatics of cultural contact.

19 S. Topik and W.G. Clarence-Smith, 'Introduction: Coffee and Global Development', in *The Global Coffee Economy in Africa, Asia and Latin America, 1500–1989*, edited by W.G. Clarence-Smith and S. Topik (Cambridge: Cambridge University Press, 2003), 2.

has incorporated a rural dimension, it has tended to focus on mapping or census taking. In contrast to this very distanced role of the state, this book looks at the everyday practices of creating and contesting legislation as it happened year by year in the colony, showing the effects of this legislation on the lives and deaths of local populations, the immigrant labourers and British plantation owners. In other words, this book is intended to contribute to the understanding of the lived experience of colonial spaces. Estates were sites where new methods were sought to manage, discipline, and regulate populations of migrant labourers by guiding and encouraging modern and efficient modes of conduct and by applying scientific methods of delivering health and welfare. One could even argue that the regulation and improvement of the estates and the lives of migrant labourers became a principal raison d'être of the colonial state, albeit one that was often contested by those whose personal economic interests lay in merely exploiting the labour of Indian migrants.

By looking at the role of the expanding plantation economy and the power of individual capitalists in shaping colonial biopower, this book demonstrates that governmentality in Ceylon had an important non-state dimension. In fact, the story of the coffee industry is a story of the clash of sovereignties and defacto sovereignties in the decision-making that decided the moral acceptability of death rates and sought to closely manage the health of the immigrant labouring population on Ceylon's plantations. Authoritarian, liberal and humanitarian forms of governmentality were worked out in the highly charged atmosphere of colonial administrators, scientists and entrepreneurs trying unsuccessfully to save an internationally important industry. Having said this, I also show that because the plantation economy was highly interconnected across global space, local decision-making on the part of government administrators and individual capitalists was often overshadowed by events elsewhere within and even beyond the British empire.

The book also adds an ecological dimension to the study of colonial governmentality, exploring, often failed, attempts on the part colonial administrators, scientists and plantation managers to manage a series of human and plant diseases that were brought into being by the radical transformation of the ecology and disease environments of the highland interior of Ceylon.[20] Where the British planters saw tropical nature turning its fury against all their hard work and dreams of future profit, botanists were beginning to discover that the disaster was the result of a complex of humans and nature (socio-nature): the economic demands of plantation agriculture, a lack of knowledge about the tropical environment and the effects of monoculture, together with a fungus inadvertently brought from Africa that produced an environmental disaster with devastating consequences for European planters, Tamil labourers and Sinhalese peasants alike. Coffee production caused the destabilization of subsistence farming in the highlands, the aggressive re-territorialization of populations from India to Ceylon, and resulted in the perceived need for the public management of the lives and deaths of labourers and peasants by colonial administrators and planters. I trace the material consequences of this ecological disaster through each of these groups, showing the differential responses of the state in each case.

---

20 P. Rutherford, 'The Entry of Life into History', in *Discourses of the Environment*, edited by E. Darier (Oxford: Blackwell, 1998), 37–61.

There are a number of nineteenth century discourses that fundamentally shaped and were shaped by the practices of colonialism in Ceylon. In the chapters to follow, I will be discussing how the discourses of the tropics, racial difference, and the arts of government helped to constitute the practices of agriculture, labour, public health, policing, discipline and plant disease management. But before moving on, I will briefly describe the discourses themselves.

Discourses of the Tropics

As Driver and Martins point out, whether conceived as a paradise of abundance or a hell on earth of disease and degeneration, the tropics have persistently served as a foil to temperate Europe and 'all that is modest, civilised, cultivated.'[21] At the heart of the European dream of transforming the tropics into viable colonies was the theory of environmental determinism, a theory that, as Livingstone[22] argues, attributed a morality to climate and led to moralistic judgements about peoples and places.[23] Theories of environmental determinism, which can be traced back to Hippocrates, assume that human behaviour is controlled to a greater or lesser extent by the environment.[24] Nineteenth century biological science modified the theory as it had been propounded by Montesquieu and others during the eighteenth century.[25] For example, the enervating heat and humidity of the tropics that had long been assumed to produce sensuality and lack of ambition now became contrasted with temperate regions more precisely in terms of climatically-induced increases in brain activity.[26] In his influential treatise *The Influence of Tropical Climates*, Johnson asserted that each race is the unique product of a particular environment.[27] And even within race,

---

21 F. Driver and L. Martins 'Uses and Visions of the Tropical World', in *Tropical Visions in an Age of Empire*, edited by F. Driver and L. Martins (Chicago: University of Chicago Press, 2005), 3.

22 D.N. Livingstone, 'The Moral Discourse of Climate: Historical Considerations on Race, Place and Virtue', *Journal of Historical Geography* 17 (1991): 413–34; D.N. Livingstone, 'Tropical Climate and Moral Hygiene: the Anatomy of a Victorian Debate', *British Journal of the History of Science* 32 (1999): 93–110; D.N. Livingstone, 'Race, Space and Moral Climatology: Notes Towards a Genealogy', *Journal of Historical Geography* 28 (2002): 159–80.

23 N. Perera, *Society and Space: Colonialism, Nationalism and Postcolonial Identity in Sri Lanka* (Boulder: Westview, 1998), 73 argues that Europeans created a 'climatic other'.

24 C. Glacken, *Traces on the Rhodian Shore* (Berkeley: University of California, 1967). D. Arnold, *The Problem of Nature: Environment, Culture and European Expansion* (Oxford: Blackwell, 1996), 19–38.

25 Livingstone, "The moral"; D.N. Livingstone, *The Geographical Tradition* (Oxford: Blackwell, 1992); Livingstone, "Tropical climate."

26 J. Hunt, 'On Ethno-Climatology; or the Acclimatization of Man', *Transactions of the Ethnological Society of London* n.s.2 (1863), 53.

27 J. Johnson, *The Influence of Tropical Climates, More Especially of the Climate of India, On European Constitutions; and the Principal Effects and Diseases Thereby Induced, their Prevention and Removal, and the Means of Preserving Health in Hot Climates Rendered Obvious to Europeans of Every Capacity*. 2nd Edn (London: B. & T. Kite, 1815), 3–4, cited in

moral climatology was fine-tuned down to the sub-regional level such that up river or at higher altitudes one found members of the same race with climate-induced mental and moral capacities and corresponding behavioural differences.[28]

While the tropical environment was thought to be unhealthy for people of all races, Europeans were seen as especially vulnerable. As Mair[29] wrote in the 1870s,

> the young man proceeding to the East, was expected as a matter of course to return home, if ever he did return, a sallow, yellow coloured, emaciated invalid, with his liver sadly damaged, his mental energies and nervous system much enfeebled, and his constitution generally so shattered, as to render him unfit for any social intercourse or enjoyment.

Mair drew upon the medical belief that the tropics produced abdominal and mental disorders, while the temperate climates of Europe were more likely to bring on thoracic disease.[30] In the early nineteenth century the humoural theory of disease had been recast in terms of tropical heat disrupting patterns of circulation and nutrition, making Europeans particularly prone to problems of the stomach and liver.[31]

Nineteenth century racism backed by the recently refined scientific ideas of environmental determinism led to a concern that people from cool regions, such as north-western Europe, would degenerate physically, psychologically, and morally if they remained in the tropics for too long.[32] It was felt that the stress of the hot sun 'unnaturally and prematurely accelerated' the maturation of the body and led to organic decline.[33] And although in fact many Europeans sought the relative coolness

---

M. Harrison, *Public Health in British India: Anglo-Indian Preventive Medicine, 1859–1914* (Cambridge, Cambridge University Press, 1994), 44.

28 Livingstone, *The Geographical*, 221–224.

29 R.S. Mair, *A Medical Guide for Anglo-Indians. The European in India, or Anglo-Indian's Vade-Mecum. A Handbook of Useful and Practical Information for those Proceeding to or Residing in the East Indies* (London: King, 1871), 213.

30 J.R. Martin, *The Influence of Tropical Climates on European Constitutions, Including Practical Observations on the Nature and Treatment of the Diseases of Europeans on Their Return from Tropical Climates* (London: Churchill, 1856), 454–55. Such a view was based upon some empirical evidence as during the first third of the nineteenth century more than half of all deaths in Britain were caused by diseases of the lung, mainly tuberculosis and pneumonia, whereas in Britain's tropical colonies they were not more than ten per cent. In Madras waterborne diseases of the stomach, liver and intestine accounted for 64 per cent of deaths of Europeans, in Britain such diseases accounted for 19 per cent of deaths (P.D. Curtin, *Death By Migration: Europe's Encounter with the Tropical World in the Nineteenth Century* (Cambridge: Cambridge University Press, 1989), 30, 35.

31 M. Worboys, 'Germs, Malaria and the Invention of Mansonian Tropical Medicine: From "Diseases in the Tropics" to "Tropical Diseases."', in *Warm Climates and Western Medicine*, edited by D. Arnold. (Amsterdam: Rodopi, 1996), 183.

32 On such beliefs in French Indo-China and the Dutch East Indies, see A.L. Stoler, 'Sexual Affronts and Racial Frontiers: European Identities and the Cultural Politics of Exclusion in Colonial Southeast Asia', in *Tensions of Empire: Colonial Cultures in a Bourgeois World*, edited by F. Cooper and A.L. Stoler. (Berkeley: University of California Press, 1997), 213.

33 Martin, *The Influence of Tropical*, 464.

of the hills in South Asia as a healthy respite from the heat of the lowlands,[34] others questioned whether tropical coolness was the same as temperate coolness. They believed that cool temperatures in the tropics were produced by humidity and great fluctuations of temperature, both of which were thought to have ill effects on the health of Europeans.[35]

While it had been the hoped that Europeans could acclimatize to the tropics, by the mid-nineteenth century few believed this possible. As Mair wrote,

> much has been said and written about acclimatizing the European in India; but the idea ... is scarcely ever seriously entertained now-a-days by those who have carefully studied the influence of the climate generally. The number of those who do not return to Europe with their constitutions unimpaired, after too long residence in India is but small.[36]

Anxieties about the possibility of successful white settlement in the tropics weighed heavily on the minds of imperialists. For example, doubts were expressed that Europeans could reproduce in India beyond two or three generations, without losing certain racial characteristics.[37] As Twining wrote, 'all the inquiries which I have been able to make, afford no evidence that the third generation of pure European descent, exists in India.[38] Likewise, Burton believed that Portuguese settlers in Goa had degenerated over the generations. He wrote that 'there is no mixture of blood, still has been one of air and climate, which comes to the same thing.'[39] In fact, the heat of the tropics was thought to change the blood of a European bringing on tropical anaemia, a condition of fatigue, listlessness and heightened susceptibility to disease.[40]

---

34 A.D. King, *Urban Colonial Development* (London: Routledge, 1976); J.T. Kenny, 'Climate, Race, and Imperial Authority: The Symbolic Landscape of the British Hill Station in India', *Annals, Association of American Geographers*, 85 (1995): 694–714; J.T. Kenny 'Claiming the High Ground: Theories of Imperial Authority and the British Hill Stations in India', *Political Geography* 16 (1997): 117–39; D. Kennedy, 'The Perils of the Mid Day Sun: Climatic Anxieties in the Colonial Tropics', in *Imperialism and the Natural World* edited by J.M. MacKenzie, 118–40. (Manchester: Manchester University, 1990); D. Kennedy, *The Magic Mountains: Hill Stations and the British Raj* (Berkeley: University of California Press, 1996).

35 Arnold, *The Problem*, 152.

36 Mair, 'A Medical Guide', 216, quoted in A. Bewell, *Romanticism and Colonial Disease* (Baltimore: Johns Hopkins University Press, 1999), 280.

37 D. Arnold, *Colonizing the Body: State Medicine and Epidemic Disease in Nineteenth Century India* (Berkeley: University of California Press, 1993), 39; M. Harrison, *Climates and Constitutions: Health, Race, Environment and British Imperialism in India, 1600–1850* (New Delhi: Oxford University Press, 1999), 19.

38 W. Twining, 'Clinical Illustrations of the More Important Diseases of Bengal, with the Result of an Inquiry into Their Pathology and Treatment' (Calcutta: Baptist Mission, 1832), 29.

39 R. Burton, *Goa and the Blue Mountains: or, Six Months of Sick Leave* (London: R. Bentley, 1851), 89. Cited in Harrison, *Climates*, 125.

40 D.H. Cullimore, 'The Book of Climates: Acclimatisation; Climatic Disease; Health Resorts and Mineral Springs; Sea Sickness; and Sea Bathing' (London: Bentley, 1890), 5.

But the idea of 'Europeanness,' if not entirely fluid, was highly complex and relational due in large part to developments in understandings of colonial Others and varying environments. Such complexity loomed large in the literature on acclimatization and medicine. For example, although Jeffreys in *The British Army in India*, worried about acclimatization, he believed it possible that those few Britons who happened to be born with a 'semi-tropical constitution', i.e., a Mediterranean appearance, could in fact 'endure long exposure to the sun in a tropical continent without serious damage to the constitution.'[41] In a similar vein, Sir William Moore, Honorary Surgeon to the Viceroy wrote that only Europeans of 'sanguine temperament' were suited to life in the tropics. The term sanguine at that time referred to a dark-ruddy complexion coupled with a passionate, irrational temperament.[42] As such, sanguine Europeans were thought to be like natives in certain crucial respects. He continued,

> theoretically, it would seem possible that the European, who in type and temperament most closely resembles the condition to which climate and mode of life has converted the native of India, would be best-fitted to encounter the adverse influences of a tropical climate; and practically this appears to be the case.[43]

Finally, until the end of the nineteenth century, it was believed by many medical experts that darker complexioned Britons were less susceptible to enteric fever (typhoid) than lighter-complexioned Britons, 'because of their physical resemblance to indigenous peoples.'[44]

Women, as the 'weaker sex,' and children were thought to succumb even sooner than men to the tropical climate.[45] The planter Millie[46] was certainly of the opinion that European women could not survive for long in highland Ceylon. The climate took its toll on Europeans in a number of different ways. Again, Millie[47] writes that 'the climate of Ceylon has a strong action on the nervous system: after a long residence the nerves get unstrung, one easily gets irritated.' Climatically-produced

---

Cited in E.M. Collingham, *Imperial Bodies* (Cambridge: Polity, 2001), 177.

41 J. Jeffreys, 'The British Army in India: Its Preservation by an Appropriate Clothing, Housing, Location, Recreative Employment, and Hopeful Encouragement of the Troops' (London: Adamant, 2005 [1858]), 1–3, quoted in Harrison, *Public Health*, 49.

42 The notion of the sanguine temperament emanates from humoural theory. It is the body type in which the element blood predominates.

43 W. Moore, 'The Constitutional Requirements for Tropical Climates, with Special Reference to Temperaments', *Transactions of the Epidemiological Society* 4 (1884–85), 37–8, quoted in Harrison, *Public Health*, 51.

44 Harrison, *Public Health*, 58.

45 Arnold, *Colonizing*, 37. Prior to the nineteenth century opinion was divided about whether women could survive the tropics better than men. This changed during the nineteenth century based more on an increased feeling by men that women were weak and should be protected than upon any empirical evidence that they succumbed more readily to disease. (Harrison, *Climates*, 90.)

46 P.D. Millie, *Thirty Years Ago: or Reminiscences of the Early Days of Coffee Planting in Ceylon*, Reprinted from the *Ceylon Observer* (Colombo: A.M. and J. Ferguson, 1878).

47 Ibid.

anxiety was thought to have a more pronounced effect on women than men, leading to various neurological disorders, including madness. Such ideas were formalised by Woodruff in the late nineteenth century in the theory of tropical neurasthenia which explained why whites failed to acclimatise.[48] Children were thought to be particularly susceptible to the ill-effects of the tropics. Lady Emily Metcalfe writing of the 1830s and 1840s warned that a mother might expect to lose three of every five children she bore in India.[49] The dangers were thought so great that even if they survived,

> [c]hildren born of white or European parents in India require to be sent to Europe in order to attain due maturity and strength. If allowed to remain in India, they seldom present the appearance of health, even when they arrive at puberty. A greater proportion of them also die before they reach this epoch of existence: and it seems probable that children, whose parents have both been the offspring of Europeans, but born and constantly resident in India, would be still weaker and less likely to arrive at maturity, or to reach the full physical development of the white variety of the species.[50]

A half century later, Davidson wrote:

> The child must be sent to England, or it will deteriorate physically and morally, – physically, because it will grow up slight, weedy, and delicate, over-precocious it may be, and with a general feebleness ... Morally, because he learns from his surroundings much that is undesirable, and has a tendency to become deceitful and vain, indisposed to study, and to a great extent unfitted to do so.[51]

Nevertheless, Arnold argues that in spite of concerns for the mental and physical vigour of Europeans living in the tropics, environmental determinism played an important function in legitimating imperialism.[52] As South Asia was considered 'subject to nature to a far greater degree than Europe, the great advantage of British rule ... was to bring about "improvement," "order" and "progress" and so liberate Indians from their servitude to nature.' Fresh blood and ideas from Britain were seen as necessary to carry out this vast imperial project, however. According to the logic of environmental degeneracy it was considered risky to leave colonial administration in the hands of long-term settlers. Consequently, improving the circuits of knowledge production between Europe and South Asia was considered to be crucial.

Discourses of Race

Although it is difficult to pinpoint exactly when ideas of racialized difference came into prominence, it is reasonable to think that the relevant genealogy was largely

---

48 Livingstone, *The Geographical*, 232–41; Kennedy, 'The perils.'

49 Harrison, *Public Health*, 50.

50 J. Annesley, *Researches into the Causes, Nature, and Treatment of the More Prevalent Diseases of India and the Warm Climates Generally* (London: Longman, Brown, Green and Longman, 1841), 43–44n.

51 A. Davidson, *Hygiene and Diseases of Warm Climates* (Edinburgh: Pentland, 1893), 5.

52 Arnold, *The Problem*, 174.

coincident with that of the bio-political state.[53] The great botanical taxonomist Carolus Linnaeus, who determined that the world of nature should be minutely classified, established in 1735 a scientific classification of humans based on skin colour. He identified four families of man, each with their own moral and intellectual particularities. His work served as a model for subsequent hierarchical classificatory schemas of race. Various taxonomists developed somewhat different systems of ranking. For example, in 1781 Blumenbach identified five races based on such criteria as skin colour, hair type, shape and size of skull and facial characteristics. Others such as Camper thought facial angle was the key discriminator of race. To these varying indicators were added nineteenth century phrenology and cranial capacity measures, all of which helped constitute an unstable brew of scientific measures of racial difference.[54] The application of science to measuring and standardising bodies according to a grid of race was but one aspect of governmentality which Scott[55] refers to as 'legibility' and 'simplification,' whereby the state ordered and controlled its subjects. We know today that the concept of race was an effect of power and a strategy of rule, to use Dirk's language.[56] Stoler elaborates the point arguing that in the nineteenth century:

> race becomes the organizing grammar of an imperial order in which modernity, the civilizing mission and the "measure of man" were framed. And with it, "culture" was harnessed to do more specific political work; not only to mark difference, but to rationalize the hierarchies of privilege and profit, to consolidate the labour regimes of expanding capitalism, to provide the psychological scaffolding for the exploitive structures of colonial rule.[57]

However, because the moral loading that racist ideas have today is so different from that of the Victorian era, it is difficult to read nineteenth century statements on the subject without projecting back the moral approbation that one feels. But if one is to understand Victorian mentalities, it needs to be understood that racism was the common-sense view of educated people who considered themselves enlightened. Furthermore, their views were seen as supported by sound empirical observations from the sciences of phrenology, anthropology, geography, and other types of

---

53 Stoler, *Carnal Knowledge*, 147.

54 J.S. Haller, *Outcasts From Evolution: Scientific Attitudes of Racial Inferiority 1859–1900* (Carbondale: Southern Illinois University Press, 1995), 4–16. It is important to note that just as things like cranial capacity were used to argue against the equality of people of colour, so in the nineteenth century these same specious indicators were used to argue that European women were incapable of rational thought. See K. Kern, 'Gray Matters: Brains, Identities and Natural Rights', in *The Social and Political Body*, edited by T. Schatzki and W. Natter. (London: Guilford, 1996), 104.

55 J. Scott, *Seeing Like a State: How Certain Schemes to Improve the Human Condition Have Failed* (New Haven: Yale University Press, 1998), 2.

56 N.B. Dirks, *Castes of Mind: Colonialism and the Making of Modern India* (Princeton: Princeton University Press, 2001), 303.

57 A.L. Stoler, *Race and the Education of Desire: Foucault's History of Sexuality and the Colonial Order of Things* (Durham: Duke University Press, 1995), 27.

colonial knowledge.[58] Having said this, in the nineteenth century race was clearly a chaotic concept. It was used not only to distinguish Europeans from others based on skin colour, but also to designate members of non-Christian religions, and various nationalities and ethnicities. The idea of race was further complicated and destabilized after the 1860s by the growing impact of Darwin's work, as racial development came to be seen as relatively fluid in evolutionary terms.[59] And to muddy matters still further, Stepan contends that race and gender were often used metaphorically, so that women sometimes occupied a quasi-racialized category and non-whites were feminized.[60] Similarly class divisions within Britain were racialized by the bourgeoisie who saw the poor as a 'race apart.'[61] For example, Mayhew argued that English society was divided into two races 'the wanderers (the urban poor) and the settlers.'[62]

Michael Adas contends that Victorians were often more concerned to discriminate difference on the basis of the technological achievement rather than on innate and unchanging racial capabilities.[63] This would seem to imply that colonialists were apt to think of racial difference in performative rather than essentialist terms, and that their judgements were based on perceived levels of modernity. He further contends that colonialists in different parts of the world operated with variant notions of race and that essentialist notions of innate racial difference were more likely to be invoked with regard to sub-Saharan Africa than places such as South and East Asia. However, Harrison argues that essentialist notions of racial difference began to gain ground in early nineteenth century India as the prospects of Europeans being able to acclimatize to the tropical climate there seemed to recede. As he goes on to point out, such racialized thinking was a double-edged sword, for although it placed Europeans in a fundamentally superior position to Indians, it also made long-term occupation of India by Europeans appear impossible.[64]

Let us now turn to a consideration of how these general conceptions of race were applied to South Asia. Late eighteenth century British Orientalist scholars like Sir William Jones, although full of praise for ancient Indian scientific and philosophical thought, believed that Indian civilization had greatly declined from its former glory. Jones argued that in regard to the sciences, contemporary Asians were like 'mere children' in comparison to Europeans.[65] During these same decades Pierre

---

58 P. Robb (ed.), *The Concept of Race in South Asia* (Delhi: Oxford University Press, 1995), 2.

59 M. Adas, *Machines as the Measure of Men: Science, Technology, and Ideologies of Western Dominance* (Ithaca: Cornell University Press, 1989), 292.

60 N.L. Stepan, 'Race and Gender: the Role of Analogy in Science', in *Anatomy of Racism*, edited by D. Goldberg (Minneapolis: University of Minnesota, 1990), 43.

61 C. Hall, *Civilising Subjects: Metropole and Colony in the English Imagination, 1830–1867* (Cambridge: Polity, 2002), 17.

62 H. Mayhew, *London Labour and the London Poor. Volume 1* (New York: A.M. Kelley, 1967 [1851]), 2.

63 Adas, *Machines*.

64 Harrison, *Climates*, 216–19.

65 Z. Baber, *The Science of Empire: Scientific Knowledge, Civilization and Colonial Rule in India* (Albany: SUNY Press, 1996). Victorians retained this comparison of Asians and

Sonnerat accounted for what he saw as India's decline from its former greatness in terms of despotism, excessive respect for tradition, and climate. The first two were thought to discourage innovation while tropical climate was held to rob people of energy.[66] Throughout the nineteenth century in South Asia, the British continued to think of Indians in this way while fearing that they themselves might decline as well under the twin impresses of a negative Indian natural and cultural environment. Such worries demonstrate how fragile and unstable differences between colonizers and colonised were thought to be. Nowhere was this clearer than in the case of the European populations in places like Ceylon, about which we will have more to say shortly.

James Mill's *History of British India* published in 1817 was a central text in shaping British opinion on India in the nineteenth century. It was required reading for servants of the East India Company and used as a textbook for candidates for the Indian Civil Service.[67] Mill was a fervent utilitarian, who viewed India and Indians in highly negative terms. Indian society, he argued, was similar to medieval Europe in many respects, and contemporary Indians' lack of scientific knowledge was proof of their inferiority to modern Europeans. Such views became widespread among the middle and literate working classes in Britain in the nineteenth century. Members of the Victorian middle class had witnessed a major transformation take place in Britain and contrasted this with the technological and social stagnation they perceived in places like India.[68] Having said this, it was thought debatable by many well-educated people in Britain that the British lower classes had been successfully transformed into modern capitalist workers.[69] As Adas points out,

> the strictures that European travellers and missionaries levelled at Africans and Asians for their indolence, improvidence, and disregard for punctuality were applied as readily by middle-class authors to the working classes, peasants, and entire "racial groups" [such as the Irish] within Europe itself ... The civilizing mission, then, was more than just an ideology of colonialism beyond Europe. It was the product of a radically new way of looking at the world and organizing human societies.[70]

For example, Mayhew and Binny in their *Criminal Prisons of London and Scenes of Prison Life*, written in 1862 referred to 'our criminal tribes' and 'that portion

---

children. They also often likened the lower classes in Britain to children on the grounds that they had weak powers of reasoning. M.J. Wiener, *Reconstructing the Criminal: Culture, Law and Policy in England, 1830–1914* (Cambridge: Cambridge University Press, 1990), 95.

66 Adas, *Machines*, 99–100.

67 Adas, *Machines*, 171.

68 A.L. Stoler and F. Cooper 'Between Metropole and Colony: Rethinking a Research Agenda', in *Tensions of Empire: Colonial Cultures in a Bourgeois World*, edited by F. Cooper and A.L. Stoler (Berkeley: University of California Press, 1997), 3.

69 On debates over the moral and biological nature of the lower classes in Britain see Stoler and Cooper 'Between metropole and colony', 12; G. Stedman Jones, *Outcast London: A Study in the Relationship between Classes in Victorian Society* (Oxford: Oxford University Press, 1971); N. Stepan, *The Idea of Race in Science, Great Britain, 1800–1960* (London: Macmillan, 1982).

70 Adas, *Machines*, 194, 153, 98, 208–09.

of that society who have not yet conformed to civilized habits.'[71] Just as liberal capitalism and utilitarianism were applied at home as central pillars of civilization, so in the colonies capitalist labour training practices and utilitarianism were thought to civilize natives, defining their own interests better than they themselves could.[72] The practices of colonial utilitarianism as part and parcel of the civilizing mission were thus based on European notions of irrational, despotic and indolent natives. As Metcalf[73]points out, India became an important laboratory in which to develop a liberal administrative state, in large part because as a conquered people they were less able to resist 'policies put in place for their own benefit' than were people in Britain.

In his medical topography and ethnography of Ceylon written in 1846 Henry Marshall, identifies five different classes of the population based on a range of different criteria. The first class is the Sinhalese, divided into Kandyans and lowlanders who share the same customs and 'like the inhabitants of the peninsula of India, have European features, the colour of their skin varying from brown to black.'[74] He cites the accounts of the moral qualities of the Kandyans by Robert Knox, a prisoner of the Kandyan king during the late seventeenth century as follows: grave and stately, crafty, unfaithful, and slothful. He justifies drawing upon this 180-year-old account in classically Orientalist terms: 'Oriental nations are very slow in changing their habits, or modes of thinking and acting.'[75] Marshall goes on to say that he considers the Kandyans to be superior to the lowlanders in 'physical force and mental energy' but that the Kandyans [like the lowlanders] are 'deficient in the principal elements of a high degree of civilization, such as wealth, truth and moral principle…'.[76] To this he adds the caveat that the differences between the present day Kandyans and lowland Sinhalese are not inherent in that, '[t]he physical and moral qualities which distinguish the Highland from the Lowland Sinhalese, it is presumed, did not exist in any remarkable degree while the Island was under one government or till after the maritime provinces had been conquered and held by a European power.'[77] The second class, the Tamils, he describes in similarly cultural terms as 'followers of Brahma' from India. The third are the Moors, whom he says are thought to be descended from Arab traders and 'may be looked upon as the most industrious and laborious class of the population.'[78] The fourth class are the Veddahs, an aboriginal population who live 'in an unsocial savage state.'[79] And the fifth class

---

71 H. Mayhew and J. Binny *The Criminal Prisons of London and Scenes of Prison Life* (London: Griffin, Bohn and Company, 1862), 384.

72 T.R. Metcalf, *Ideologies of the Raj* (Cambridge: Cambridge University Press, 1995); R. Iyer, *Utilitarianism and All That* (London: Concord Grove Press, 1983), 11–13.

73 Metcalf, *Ideologies*, 29.

74 H. Marshall, *Ceylon: A General Description of the Island and its Inhabitants* (London: H. Allen, 1846), 13.

75 Ibid., 14.

76 Ibid., 15.

77 Ibid., 15.

78 Ibid., 13.

79 Victorians distinguished between 'savages' or 'primitives,' such as tribal peoples living at a very low level of material accomplishment and 'barbarians' whose societies were

are the Burghers 'descendants of Europeans of unmixed blood, and that race which has sprung from the intercourse of Europeans with the natives.'[80]

Marshall's descriptions exemplify the chaotic conceptions of race current at that time. Although he did not consider any of the Sinhalese to be civilized, he prefers the 'noble' highlanders to the lowlanders whom he considers to be corrupted by several centuries of Portuguese and Dutch rule. And yet while acknowledging the impact of cultural change through interaction with Europeans; he maintains the view of Ceylon as essentially unchanging. A persistent stereotype that pervades the images of all the natives of the island held by Marshall and others is of physical weakness and femininity. Lewis, for example, characterises the Kandyans as lazy and apathetic and expresses surprise that the low country Sinhalese, although a 'weaker race,' are better workers than the Kandyan highlanders.[81] His surprise stemmed from the widely held belief that people from cooler regions and even sub regions are naturally superior.

Most of the labour on the plantations, however, was performed by Tamil labourers imported from India. Whilst they were not mentioned in Marshall's text, as few came to the island when he wrote, they were later typically represented as: 'docile, indisposed for exertion, but with great powers of endurance, influenced in their conduct by fear or by love of gain, rather than by motives of gratitude or a sense of duty ... Their habits are purely sensual. The indulgence of their animal appetites is their sole motive for exertion.'[82] A recurrent theme in descriptions of all the islanders was delicacy and effeminacy. Jenkins[83] spoke in thinly veiled homoerotic terms of how on his arrival he was struck by the sight of 'naked coolys, men who seemed to be all women ...' They were seen in stark contrast to the moral and physical masculinity of the European.[84] Some authors such as Harrison suggest that effeminacy was thought to characterize not just South Asians, but all tropical people.[85] Thomas, however, believes that whilst effeminacy was particularly associated with Asians, it was not in the least with other tropical people such as Melanesians. But such characteristics were not seen as immutable, and many believed that the power of European capitalism could counteract climate and teach natives to be more manly. As the colonial surgeon Dr. Van Dort[86] put it, 'the genius of labour ... stamps the physiognomy [of the Tamil labourer] with an expression of manliness and intelligence which is never seen in

---

thought to be in decline. The Chinese and Indians were thought to be examples of the latter. Adas, *Machines*, 194–95.

80 Marshall, *Ceylon*, 14.

81 R.E. Lewis, *Coffee Planting in Ceylon, Past and Present* (Colombo: Examiners Office, 1855), 9–10.

82 *Proceedings of the Planters' Association (Kandy)*, (henceforth *PPA*) 1869–70, 56.

83 A. Planter (Richard Wade Jenkins) *Ceylon in the Fifties and the Eighties, a Retrospect and Contrast of the Vicissitudes of the Planting Enterprise During a Period of Thirty Years and of Life and Work in Ceylon* (Colombo: A.M. and J. Ferguson, 1886), 1.

84 For a discussion of Indian effeminacy, see Metcalf *Ideologies*, 9.

85 Harrison, *Climates*, 59. Thomas, *Colonialism's Culture*, 133.

86 *PPA*, 1869–70, 55.

the raw, uncivilized, newly landed cooly.' It would transform him, in the words of an Assistant Government Agent (AGA) into a 'lusty labourer.'[87]

By the late nineteenth century, such mid-century views as those of Marshall and others had not fundamentally changed. The Kandyans continued to be seen as lazy and crafty, but with a certain nobility to them. The lowland Sinhalese, on the other hand, especially those who migrated to the central highlands in search of employment, increasingly came to be thought of as a criminal class. As Rogers points out, the British applied the same social categories to the Sinhalese as they did to the under classes in Britain.[88] Hence, the lowland Sinhalese corresponded to the Victorian 'criminal classes' in Britain and the Kandyan peasantry fulfilled the role of 'the respectable poor.'[89] But he adds, cross-cutting this was the Anglo-Indian concept of 'criminal castes and tribes' who were thought to have innate criminal tendencies, brought on, in the case of the lowland Sinhalese, by a luxuriant environment that breeds laziness and criminality.[90] We will see how these stereotypes shaped the interactions between the British and those they encountered in Ceylon in subsequent chapters.

Perhaps Marshall's most interesting commentary concerns the Burghers, a category that includes descendants of Europeans and those of mixed race. This is most revealing of the complexity of colonial categories and undermines any simple white/brown or other dualism. The category 'Burgher' is important taxonomically because it includes both white Europeans of Portuguese and Dutch descent and mixed race islanders. Such a breaching of the racial divide shows that European Burgers were considered a relict population that had degenerated under the impress of the natural and social environment and was thus no longer considered European. As Stoler[91] points out with reference to the Dutch East Indies, this type of categorization raises the interesting issue of just what it meant to be white and European.

Roberts et al., argue that although the category of Burgher includes Dutch descendants, Dutch/native mixes and Portuguese/native mixes (known locally as tupass), those Burghers who claimed 'pure' Dutch ancestry in fact looked down on the other two categories and the British shared this view. Although the upper levels of Burgher society were brought into junior administrative positions within the colonial government, they were seen as socially inferior to the British. While the 'pure' Dutch Burghers looked down on other darker-skinned Burghers, by the 1840s the racial hardening of British society increasingly cast all Burghers as 'native,' as

---

87 *Administration Reports, Ceylon* (henceforth *ARC*) 1867, 40.

88 On the way in which colonial categories were in turn used to talk about the poor in Britain, see S. Thorne, '"The Conversion of Englishmen and the Conversion of the World Inseparable": Missionary Imperialism and the Language of Class in Early Industrial Britain', in *Tensions of Empire: Colonial Cultures in a Bourgeois World*, edited by F. Cooper and A.L. Stoler, 238–62. (Berkeley: University of California Press, 1997).

89 J.D. Rogers, *Crime, Justice and Society in Colonial Sri Lanka* (London: Curzon Press, 1987), 210–11.

90 Ibid., 211. On London's poor as criminal tribes see Mayhew and Binny, T*he Criminal Prisons*, 384–86, who define the 'criminal character' as marked by a love of ease and repugnance for hard work.

91 Stoler, *Race*, 114.

hybrids that were in some respects inferior to 'pure' natives.[92] For example, Forbes[93] writes of the fatal drop of native blood that leads inevitably away from whiteness:

> The slightest mixture of native blood with European can never be eradicated and in some cases seems to go on darkening in each succeeding generation, until as in many of the Portuguese descendants; we find European features with jet-black complexions. The Dutch descendants, with native blood, are now undergoing the blackening process, although they have only reached as far as a dark and dingy yellow.

Sirr[94] articulates a typically racist view that racial mixing weakens both races:

> [t]he half castes of Ceylon or Burghers as they are called in the island, adopt the European costume. ...The male half castes are far below the Cingalese both in physical power, stature, personal appearance, and mental capabilities; their complexions are less clear, their features ill formed, and the expression of their countenances is heavy and sensual, being as deficient in corporeal attractions as they are destitute of moral rectitude and probity. The females of this class in early life are remarkable for their beauty, but all traces of which are totally lost before they are thirty years of age, then they are either shapeless masses of flesh or reduced to skin and bones. It is most extraordinary, but all those who have been in the East frankly admit that among the half castes is to be found every vice that disgraces humanity, and nowhere is this axiom more strikingly exemplified than in the male and female Burghers of Ceylon.

On the other hand, another early commentator on race in Ceylon, Bennett,[95] does not share Sirr's view of Burghers arguing that '[t]he government clerks are selected from these [Burgher] families, and manage all the clerical duties of the public offices in an admirable manner:' As we can see from the above, at mid-century the British held varying views of Burghers, not all of which were negative, but all of which distinguished them from Europeans.

The issue of Europeanness is further complicated by the fact that second and third generation English settlers were considered British rather than Burgers. Although they were thought to have degenerated and were considered inferior to the more recently arrived British settlers, they had not entirely dropped out of the official category 'European.' Instead 'country-bred' or 'country bottled' were informal, pejorative terms used to describe them. While such a distinction appears nowhere in official records, it served as an informal basis for discrimination and demonstrates

---

92 M. Roberts, P. Rheem, and C. Thome, *People Inbetween: The Burghers and the Middle Class in the Transformation of Sri Lanka, 1790–1960* (Ratmalana: Sarvodaya Book Publishing, 1989). It is important to note that while according to racial theory, the mixing of races would produce offspring inferior to either race (see, Haller, *Outcasts*, 28–32), in practice the British greatly favoured mixed-race Burghers to the so-called pure native races in Ceylon. This suggests that although racial theory was important in shaping thought, such ideology was often contradicted by practice.

93 M. Forbes, *Eleven Years in Ceylon, Comprising Sketches of the Field Sports and the Natural History of that Colony, and an Account of its History and Antiquities, 2 Vols.* (London: Richard Bentley, 1840), 162–63.

94 H.C. Sirr, *Ceylon and the Cingalese Vol. 2* (London: William Sholbert, 1850), 40–41.

95 J.W. Bennett, *Ceylon and its Capabilities* (London: W.H. Allen, 1843), 157.

both how complex the colonial hierarchical systems were and how chaotic the notion of race was in terms of practice. I will explore this in more detail in Chapter 3.

Discourses of Governmentality and Biopower

Foucault puts forth the intriguing proposition that eighteenth century Europeans 'invented' what he calls 'a synaptic regime of power.' He describes this power as diffuse, as one that is exercised within individual bodies, inserting itself 'into their actions and attitudes, their discourses, learning processes and everyday lives' rather than being imposed 'from above'.[96] I do not wish to address here the question of the historical specificity of the European experience of modern forms of power working through individuals and their bodies. In fact, I would tend to agree with Comaroff and Comaroff, that 'the body ... cannot escape being a vehicle of history,'[97] However, I do follow up the idea of modern governmentality as synaptic power, a form of biopower, and explore the specific ways it did and did not apply in nineteenth century Ceylon. I explore what Scott calls the 'rationalities of colonial power,' the point of application of which was the body and 'the conditions in which that body is to live and define its life.'[98] Central to this project was the physical environment and the geographical limits to modernity; understood as 'an ambiguous zone of failure at the edge of European modernity.'[99] Now this 'zone of failure' need not be seen as geographically marginal, nor is it necessary to see modernity as centred on Europe. Indeed , the colonial experience was in many ways central to a vast entangled, highly diverse, geographically and socially uneven, multitude of projects and transformations around the world that can very loosely be called modernity. It is this 'zone of failure' that I wish to explore, but not as failure in the sense that the colonial projects in Ceylon had no significant impact; they did of course. Rather, I look at failure as 'productive failure'[100] and at qualifications of modernity and its universalizing ideals. It is important to understand, however, that the qualifications I explore are from the various points of view of nineteenth century British colonizers, not from a twenty-first century postcolonial perspective. I am interested in how difference was researched, theorized, debated, and employed to justify the practices of colonialism in Ceylon at that time. There was a failure of certain contradictory ambitions of governmentality and liberalism, but the British colonial interventions and local resistance to these interventions were necessarily productive of long-lasting environmental, political, social and economic changes in Ceylon and elsewhere.

Governmentality is a modern form of decentred pastoral control of a population conceived of as a biological phenomenon with the aim of improving welfare, habits of mind, and the conduct of everyday life including importantly self-discipline.

---

96 M. Foucault, 'Prison Talk', in *Power/Knowledge: Selected Interviews and Other Writings, 1972–1977*, ed. C. Gordon, 37–54. (New York: Pantheon, 1980), 39.

97 J.L. Comaroff and J. Comaroff, *Ethnography and the Historical Imagination* (Boulder: Westview, 1992), 79.

98 Scott, 'Colonial Governmentality', 30.

99 Redfield, 'Foucault in the Tropics', 52.

100 Ibid., 52.

According to Foucault, 'population, comes to appear above all else as the ultimate end of government. In contrast to sovereignty, government has as its purpose ... the welfare of the population, the improvement of its condition, the increase of its wealth, longevity, health, and so on.'[101] This modern, expanded view of instructing, managing, disciplining and regulating the self and others under one's control was based on the assumption that such aims could be accomplished through the displacement of traditional practices and unscientific beliefs by modern rationality. This form of decentralized power depended on the use of things like the census, surveys and various other bureaucratic technologies for monitoring populations.[102] It depended on acquiring new social scientific knowledge that could be deployed in disciplining and educating members of a population. Although the bureaucratic control over the health and welfare of populations was based on genuine humanitarian concerns, it was also considered an economic necessity.[103] Foucault argues that governmentality and the biopower upon which it depends, '... was without question an indispensable element in the development of capitalism; the latter would not have been possible without the controlled insertion of bodies into the machinery of production and the adjustment of the phenomena of population to economic processes.'[104] However, as I will show, there were different interests within the colonial world that did not always agree on the exact calculations of the costs and benefits of pastoral concern. Furthermore, there were those whose economic interests and experiences led them to argue that the racial differences of colonized populations from European-derived norms resulted in differential capacities and potentialities, and thus the management of their lives and deaths should be subject to different economic, political and even moral principles.

The science of bio-politics arose in the latter part of the eighteenth century as the systematic disciplinary knowledge of the individual and social body. Such knowledge was seen as central to the rational intervention of the state in the management of health. Austrian physician Johann Peter Franck proposed a disciplinary system of health administration which he termed 'medical police.' In a six volume thesis, he outlined methods for regulating individual behaviour (marriage, pregnancy, personal hygiene), as well as public health measures (sanitation, clean water supply, overcrowding, vice, etc.). Underpinning all of this was the notion of a population

---

101 M. Foucault, *The Essential Works of Foucault 1954–1984; Power* (London: Penguin, 2001 [1978]), 216. For a discussion of earlier forms of pastoral power and expectations of welfare by the state on the part of the Kandyan peasantry during the time of the Kandyan kings, see J.S. Duncan, *The City as Text: The Politics of Landscape Interpretation in the Kandyan Kingdom* (Cambridge: Cambridge University Press, 2005).

102 For an extended study of governmentality in an historical geographical context see M. Hannah, *Governmentality and the Mastery of Territory in Nineteenth-Century America* (Cambridge: Cambridge University Press, 2000).

103 On colonial philanthropy see: D. Lambert and A. Lester. 'Geographies of Colonial Philanthropy', *Progress in Human Geography* 28 (2004): 320–41; A. Lester, 'Obtaining the "Due Observance of Justice": The Geographies of Colonial Humanitarianism', *Environment and Planning: D* 20 (2002): 277–93.

104 M. Foucault, *The History of Sexuality: An Introduction. Volume 1* (New York: Vintage, 1990), 141.

as a resource to be managed by the state through other institutions including the family and other sites of social control and self-governance.[105] European ideas of governmentality were of course continually debated, negotiated and qualified as colonial administrators considered variables such as the finances of the colonies and of individual plantations, and, more importantly, ideas of tropicality and race as the basis for exceptions to the working out of modernity in practice.

As I will show, the discourses of governmentality were seen as very much open for negotiation and were often pragmatically adopted or rejected by representatives of different interests. Dean says that power is not played out 'within an a priori structural distribution. It is rather the (mobile and open) resultant of the loose and changing assemblage of governmental, practices and rationalities.'[106] These practices include forms of knowledge such as the changing techniques and instruments of visualizing the population to be governed.[107] Foucault describes the knowledge practices and larger discourses of which they are a part as contested and productive of changing power relations. He says that, '[t]here is not, on the one side, a discourse of power, and opposite it, another discourse that runs counter to it. Discourses are tactical elements of blocks operating in a field of force relations; there can exist different and even contradictory discourses within the same strategy.'[108] As I hope to show, the working out of the principles and practicalities of modern governmentality in the tropical colony of Ceylon was fraught with internal controversy, tactical use of theory and not a small amount of hypocrisy and self-interest. It was also met with considerable resistance on the part of the colonized populations. To quote Foucault again, '... one would have to speak of *biopower* to designate what brought life and its mechanisms into the realm of explicit calculations and made knowledge-power an agent of transformation of human life. It is not that life has been totally integrated into techniques that govern and administer it; it constantly escapes them.'[109] Of course, this escape is particularly marked where cultural differences between the administrators and the target population are great and political legitimacy is weak.

---

105 D. Porter, *Health, Civilization and the State* (London: Routledge, 1999), 52–3.

106 M. Dean, *Governmentality: Power and Rule in Modern Society* (London: Sage, 1999), 29.

107 Dean (ibid.), argues that various forms of visibility are necessary to the operation of particular regimes. He says for example that architectural drawings, management flow charts, maps, pie charts, graphs and tables are among the ways of 'visualizing fields to be governed.' Matthew Hannah (*Governmentality*) provides a well researched, in-depth study of the US census during the late nineteenth century as a large scale instrument of observation and the making of a population and its territory governable 'at a distance'. Also on observation and government see James Scott, *Seeing Like a State*. And on mapping see M. Edney, *Mapping an Empire: The Geographical Construction of British India, 1765–1843* (Chicago: University of Chicago Press, 1997). Steven Legg's *Spaces of Colonialism* offers an insightful geographical perspective on the visualization of a people and landscape by a government. Also see B. Cohn, *Colonialism and its Forms of Knowledge: The British in India* (Princeton: Princeton University Press, 1996).

108 Foucault, *The History of Sexuality: Vol. 1*, 101–102.

109 Ibid., 143.

While modern governmentality adopted as its general focus the everyday working and moral lives of the colonized populations, especially the plantation workers, the rationality of power, or what Foucault has called the micro-practices of power, in fact were worked out differently in each institutional domain. During this period, liberal principles of 'natural rights' and attempts at 'moral improvement' of the colonial populations were seen to be severely challenged by many of the British and indeed some of the non-Europeans in positions of authority. These modern liberal ideals were thought to be compromised in colonial situations because the hegemony of the British was insecure, modern European visions of civil society had not been established, and cultural and racial differences were thought to be significant to the practical working out of such ideals. Nevertheless, there arose vociferous debates among political thinkers and politicians about the extent to which a program of cultural imperialism could and should be accomplished. Among the British and non-European elites there were many different opinions on what the role of state medicine or public health in producing healthy and productive colonial subjects should be and about the range of possibilities given the differences in race and medical traditions between South Asian populations and Europeans. Many retained a view more akin to James and John Stuart Mill than to Bentham, believing that colonial rule must be despotic and paternal rather than democratic; that the perceived frailties of the South Asian body and character justified authoritarian governmentality. Theories of racial and environmental difference impacted on liberal and utilitarian philosophies of colonial government justifying oppressive measures of control as long as the end result was the improvement of a population.

Organisation of the Book

In the next chapter I discuss the coffee estates in relation to the environmental and political-economic contexts of the island of Ceylon. I also situate Ceylon coffee within the networks of world coffee production and distribution. I begin by tracing the political ecology of the highlands as the British found it at the conquest of the Kandyan kingdom in 1815. I then discuss the political, economic and environmental interventions of the British in Ceylon as these developed in response to the world demand for coffee. Chapter 3 explores the cultural anxieties of white, male planters as they attempted to retain their identities as middle-class Europeans in the face of what they saw as the degenerative influences of tropical climate and the presence of large numbers of non-Europeans. I particularly focus on moral hygiene as a form of self-disciplinary practice. Chapter 4 examines the strategies that planters employed in attempting to discipline Tamil migrant labourers and the tactics used by the labourers to evade such discipline. I further explore the implications of the coffee estates as de-facto sovereign spaces where the planters took control over the labourers as a population whose lives and rates of death were calculated in economic and moral terms. In Chapter 5 I address the bio-political responses of the colonial government to a poorly understood and frightening tropical disease environment and the weighing of various interests in this response. I also discuss different spatial strategies for the management of cholera and sanitation and the planters' critique of what they saw as improper governmental intrusion into the effectively 'sovereign' spaces of the estates. Chapter 6 explores the planters' and the state's unsuccessful attempts to make crime on the borders of the estates more visible and hence more manageable. Once again, tensions over governmentality led to great controversy in the framing of policy on crime. In Chapter 7, the final years of coffee estates on the island are documented. The chapter describes the various ways that the economic demands of plantation agriculture and lack of knowledge about the tropical environment produced an environmental disaster with devastating consequences for European planters, Tamil labourers and Sinhalese peasants alike.

Chapter 2

The Rise of a Plantation Economy

Introduction

The island of Ceylon lies forty miles across the Palk Straits from the southern tip of India. Its first human inhabitants probably arrived around 28,500 BCE occupying most of the island at very low densities. Ceylon's location relative to the Indian mainland as well as its geomorphology, varying climatic zones, and flora and fauna have been fundamental to its particular historical trajectory. Its location is deeply integrated into Sinhalese cultural and political discourses. For example, its separation from the Hindu mainland has been employed as a major justification for its political role as protector of Buddhism and this has been fundamental to Ceylon's identity and destiny. Nevertheless, the Palk Strait was sufficiently narrow that continuing migrations and political pressure from India was exerted upon the island. In fact, the beginnings of the Kandyan Kingdom in the thirteenth century can be traced from a forced move of the Sinhalese from the dry zone to the central highlands in order to escape Tamil raids from South India.

The Portuguese arrived on the coasts of Ceylon at the beginning of the sixteenth century and by a century later had annexed the coastal kingdoms leaving the Kandyan kingdom in the central mountains as the sole remaining indigenous power on the island. The Kandyans held out against repeated attacks by the Portuguese and after them the Dutch. This poor state had been protected for three centuries from European armies by the combination of an inhospitable mountainous terrain, tropical diseases that devastated European troops, and guerrilla tactics. The Kandyans won campaign after campaign not by fighting battles, but by temporarily abandoning their capital to invaders, cutting supply lines, staging ambushes, and depending on tropical nature to weaken their adversaries. While such strategies worked well in the mountain fastnesses at the heart of the kingdom, they were less successful on the flat coastal plains of the island, and consequently, the Kandyan state became increasingly impoverished as it lost its access to the sea.

Ceylon had been long established as a node in an extensive trade network linking it to southern Europe and parts of East Asia. The Portuguese and the Dutch had used Ceylon mainly for its ports and coastal regions. While the British were initially keen to develop the deep water port of Trincomalee on the east coast, they later attempted to colonize the whole of the island. The Kandyan kings had forbidden the building of roads up from the coast and deployed an army to defend the few mountain passes which existed, for it had long been a belief in the kingdom that Kandy would fall someday to an invading army that could build a road to Kandy through the jungle and miraculously pass through the middle of a mountain. After the British defeated the Kandyans in 1815 they constructed a road from the port city of Colombo to Kandy and built a tunnel through a mountain.[1] The latter effort was a conscious attempt to fulfil the myth of conquest. Thus the eventual defeat of the kingdom by the British has been termed 'a triumph of military engineering.'[2] Once the kingdom had been conquered and coffee plantations were established, the island became increasingly integrated into the world economy.

The Environmental Framework

Geomorphologically, Ceylon consists of a central mountainous core with peaks ranging from five to eight thousand feet. This central core has eroded to form a broad coastal plain. The coastal and highland soils in general are not very fertile and the fertility is uneven. The island's climate is dominated by two monsoon seasons, one in the summer drops rain on the south-western portion of the island with the highest precipitation falling in the highlands. The mountains create a rain shadow so that very little rain reaches the south-eastern portion of the island during this season. Then during the winter monsoon, north-eastern winds bring rain to the north and south-eastern portions of the island. This monsoonal and inter-monsoonal pattern of rainfall produces significant geographical variation in rainfall across the island. It creates a wet zone in the southwest surrounded by an intermediate zone. To the north and east of this lies the dry zone. The central mountains have the most precipitation with the western portion receiving between 100 and 200 inches per year and the south-eastern portion receiving 50–75 inches per year. Although there is little seasonal variation in temperature, it drops at a rate of three to four degrees Fahrenheit per 100 feet of elevation, creating a steep temperature gradient between the coastal plain and the high mountains.[3] As we will see, this spatial unevenness in altitude, rainfall, and temperature along with other historical and political factors, led to uneven rates of success in plantation agriculture during the nineteenth century.[4]

The Agricultural Economy of the Kandyan State

During the period of the Kandyan kings before the British conquered the central highlands in 1815, most settlements lay well below 3,000 feet, as rice, the staple

---

1   For a discussion of the politics of the royal capital of Kandy on the eve of the British take over, see Duncan, *The City as Text*.

2   N. Perera, *Decolonizing Ceylon: Colonialism, Nationalism, and the Politics of Space in Sri Lanka* (Oxford: Oxford University Press, 1999), 43.

3   J.L.A. Webb, *Tropical Pioneers: Human Agency and Ecological Change in the Highlands of Sri Lanka, 1800–1900* (Athens: Ohio University Press, 2002), 7–8; T. Somasekara, et al. (eds), *Arjuna's Atlas of Sri Lanka* (Dehiwala: Arjuna's Consulting 1997).

4   G.H. Peiris, *Development and Change in Sri Lanka: Geographical Perspectives* (New Delhi: Macmillan, 1996), 26–33. See Webb above, for the most complete environmental history of Ceylon to date. Webb's excellent volume explains in depth the relationship between the environment of highland Ceylon and the rise and fall of the coffee industry from an ecological point of view.

crop, was difficult to grow above this altitude because of low temperatures and heavy cloud cover.[5] The forced move during the thirteenth century of the Sinhalese into the central highlands had been not only a move from large scale irrigation agriculture to small scale paddy and burned forest land (*chena*) used for dry grain cultivation, but a move from large agricultural surplus to subsistence farming. The typical highland village consisted of lands devoted to wet rice paddies, gardens adjacent to people's houses, and forest lands for chena farming and the collection of firewood and honey. Throughout the wetter regions within the highlands, paddy cultivation was fed by seasonal rains, but in the drier eastern rain shadow, water for irrigation had to be stored in small tanks. Given the low population density of the highlands, it was labour rather than land, which was the scarce factor in production.[6]

As the owner all of the land in the kingdom, the king allowed his subjects to cultivate land in exchange for goods and services for him and his nobles.[7] Tenant cultivators had inalienable rights to use and transfer rice lands so long as they paid their rents, in the form of service or produce. As there was no wage labour in the kingdom, villagers exchanged labour during the labour-intensive growing season.

The abundant forestland surrounding villages provided a crucial supplement to the villager's paddy lands. Villagers communally cleared and burned tracts of forest and then divided them into family plots.[8] Cultivators had exclusive rights over the produce from these chena plots. A wide range of crops were grown such as maize, finger millet, dry hill rice, and various types of beans. Much of the work on this chena land was done during the slack season for wet rice cultivation. After the harvest, the land was abandoned for a period of five to 25 years. It is estimated that when the British arrived in the highlands there was approximately one acre of chena under cultivation for every two of paddy.[9] In addition to chena agriculture, villagers maintained fruit and vegetable gardens and as villages expanded, chena land became progressively incorporated into these individual gardens.[10] While paddy lands were taxed, villagers paid no tax on garden or chena land. Therefore, as Bandarage points out, chena 'represented a realm of political autonomy for the cultivator class providing the peasants a sphere wherein they could resist harsh surplus appropriation by the feudal overlords.'[11]

There were four types of villages in the kingdom: those belonging to the king, the nobles, the Buddhist temples, and free villages. Structurally these villages were very similar; the major difference being to whom the rents were due. As the kingdom was poor, with little specie in circulation, the king depended on certain villages to support the royal household. Tenants in those villages gave the king a share of the

5   Webb, *Tropical*, 31.

6   A. Bandarage, *Colonialism in Sri Lanka: The Political Economy of the Kandyan Highlands 1833–1886* (Berlin: Mouton, 1983), 19–21.

7   Ibid., 21.

8   H.W. Coderington, *Ancient Land Tenure and Revenue in Ceylon* (Colombo: Ceylon Government Press, 1938), 4.

9   A.C.M. Ameer Ali, 'Rice and Irrigation in the Twentieth Century.' *The Ceylon Historical Journal* 25 (1978), 253.

10 Bandarage *Colonialism*, 26–28.

11 Ibid., 31.

rice they produced and were liable for one to three months of *rajakariya* (corvee labour) per year. The king used land grants to reward and control the nobility or to garner political support. The right of a noble to exact tribute from these villages was at the pleasure of the king and if a noble family fell out of favour, these village lands could be removed from its control. As such, control over surplus generated by villages was the most important component of the power struggle between the royal household and the nobility.

Lands were also granted by the king to temples as a display of meritorious behaviour and patronage of religion. Peasants preferred working on temple lands as the rents they paid were normally lower than under other forms of tenancy. As grants to temples were irrevocable, aristocratic families often tried to gain security by positioning themselves within the clergy. Thus the aristocracy exerted much of their economic and political power through the control of temple villages.

Although the thick forest cover in the mountainous central highlands was a difficult environment in which to generate an agricultural surplus, the Kandyans purposely maintained the hostile nature of the environment by forbidding the cutting of this barrier forest , forbidding the building of roads and keeping the few paths that wound up from the coast narrow and in poor repair. Having said this, the kingdom had long been in contact with the wider capitalist world. Although isolating itself for protection, it managed to retain a fair measure of control over its own economy during the period that the Portuguese controlled lowland Ceylon. During that time the highlands supplied food to the lowlands and two thirds of the island's exports came from the Kandyan hills.[12]

Much of the trade within the kingdom was handled by Muslims. The major imports were cloth, salt and dried fish and the major exports were areca nuts, cinnamon, and pepper. However, trade was severely limited by a number of factors. The first was the physical difficulty of transport within the highlands. The same ill-maintained paths through the forest belt that helped keep European armies at bay also slowed trade. In the mid-eighteenth century the Dutch captured all the island's ports and henceforth all external trade was conducted on terms unfavourable to the Kandyans. Given these difficult conditions and the small size of the surplus that could be generated by peasant cultivators in such a heavily forested tropical environment, it is easy to see why the kingdom became increasingly impoverished.[13] As Bandarage states:

> It is clear that the effect of Dutch mercantilism on the Kandyan economy was forced isolation and involution, rather than economic expansion. This meant that with the decline of commerce, the money supply of the state, which had not been great to begin with, was further reduced. It meant also that the state had to pay its officials increasingly with land grants in lieu of money (even though) ... land was the basis of political authority in Kandyan society.[14]

---

12 Ibid., 44.

13 Ibid., 42–44; S. Arasaratnam, *Ceylon* (Englewood Cliffs, New Jersey: Prentice Hall, 1964).

14 Bandarage, *Colonialism*, 46.

The Political Organization of the Colonial State

The British displaced the Dutch on the coasts of Ceylon in 1796, but continued the Portuguese and Dutch style of mercantilism which utilised pre-existing forms of production and distribution and relied upon state monopolies and trade restrictions. The revenues generated from the colony were shipped back to the metropole, contributing to the accumulation of capital there. After defeating the Dutch, the British continued to exact cinnamon, the traditional export crop, using *rajakariya* (corvee labour). At first, the British encouraged trade and discouraged investment in agriculture as had the Portuguese and Dutch before them. Under mercantilism, there was very little interest in the transformation of the native social order, in fact the British ruled through traditional power hierarchies thus reinforcing them. The primary interest of the mercantilist state was revenue collection. In fact, the state collectors and their native intermediaries were so rapacious that pre-British levels of tax were greatly exceeded prompting revolts against British rule in the Maritime Provinces in 1797 and in the Kandyan provinces after 1815.

By the beginning of the 1820s mercantilism was increasingly challenged by new economic theory based in laissez-faire philosophy, both in Britain and in the colonies. This was due in part to the rise of the middle classes in Britain who favoured fresh ideas such as free trade, utilitarianism and evangelicalism.[15] Places such as South Asia were seen as crying out for reform and were considered ideal testing sites for new policies.[16] In Ceylon, the opening up of the country to investment capital for plantation agriculture was seen as an important aspect of this reform movement.[17]

Perhaps most important to the introduction of plantation enterprise in Ceylon were the Colebrooke-Cameron Reforms of 1833.[18] These, in the words of Bandarage,[19] 'provided the politico-juridical framework for the "modernization" of the island.' In contrast to Bandarage's and other typically teleological narratives of modernization, Scott describes the Colebrooke-Cameron Reforms as the 'displacement of one kind of political rationality – that of mercantilism or sovereignty – by another, that of governmentality.' He explains that under mercantilism the point had been the extraction of wealth and that as long as the local population cooperated in the mercantilist projects there was no reason for British administrators to concern themselves with the lives of the colonized population; their 'local habits,' 'ancient tenures,' 'distinctions', 'religious observations' were 'not a significant variable in the colonial calculus (at least as long as they did not interfere with the immediate business of extraction).'

The Reforms were designed to alter the colonial project so that the point of application of power became the 'society itself.' And such power was intended to respect the 'natural rights' of the people. At first it was Bentham's liberal brand of

---

15 On the role played by evangelicalism in the rise of middle class power in Britain, see Lester, 'Constituting Colonial Discourse', 31.

16 Collingham, *Imperial Bodies*, 51.

17 Bandarage, *Colonialism*, 52–58, 303.

18 G.C. Mendis (ed.), *The Colebrooke-Cameron Papers: Documents on British Colonial Policy in Ceylon 1796–1833* (Oxford: Oxford University Press, 1956).

19 Bandarage, *Colonialism*, 59.

utilitarianism, which emphasized the necessity of achieving cultural hegemony and the role of the state as a guardian of a population, rather than John Stuart Mill's more authoritarian version that underpinned these reforms. As we shall see, the main thrust of the reforms was to ensure the transition of the British administration of the island from that of direct engagement in the economy to facilitation of capitalist enterprise. Scott[20] argues that the British adopted a new strategy of government oriented towards 'improving the conditions of social life including moral conduct,' replacing an older mercantile orientation towards territorial expansion and the extraction of revenue and forced labour in the form of *rajakaryia*. Scott[21] states, '… what was at stake in the governmental redefinition and re-ordering of the colonial world was, to paraphrase Jeremy Bentham … to design institutions such that, following only their own self-interest, natives would do what they ought.' Here Scott refers to the beginnings of an era of modern colonial governmentality and its displacement of older mercantilist forms of political rationality.[22]

Although in theory the new role of the state was merely to provide a conducive environment for the development of capitalism, including importantly the reproduction of the population of migrant labourers, in actuality, as we will see, it took a strongly interventionist role whenever such enterprise appeared to be threatened.[23] The Colebrooke-Cameron Reforms established a system of government organized along utilitarian lines. Nine provinces composed of 21 districts in total were created with provincial boundaries designed to cut across traditional political and ethnic lines so that the country might be governed as a unity.[24] An administrative structure was put in place whose authority took precedence over the native administrative structure. At the apex of the colonial administration was the Governor. Colonial policy was largely made by the Governor and his Executive Council made up of senior officials. The Governor also presided over a Legislative Council composed of British colonial officials who made up a majority, a permanent minority which included Sinhalese, Tamils, and Burghers, and another three places reserved for European residents drawn from planting and commercial interests and from the general European population. While all legislation had to be approved by the Legislative Council, the Governor was assured of the support of the officials who controlled the vote. All legislation had to be confirmed by the Colonial Office. The Governor administered through the Colonial Secretary and department heads reported to the Secretary. Outside the capital, the principal administrators were the Government Agents (GAs) and Assistant Government Agents (AGAs) who had charge of provinces and districts. AGAs, in turn, administered their districts through headmen. GAs and AGAs were responsible for law and order, revenue collection and implementing land and tax law.[25]

---

20 Scott, 'Colonial Governmentality,' 23–49.
21 Ibid., 44.
22 Ibid., 37.
23 Bandarage, *Colonialism*, 63.
24 Rogers, *Crime*, 20.
25 Ibid., 18–20.

The legal system which the British instituted was similar to that of Britain; but it was more progressive in certain ways. Capital punishment, for example, was only meted out for treason and murder after 1796 in Ceylon, but was continued for other crimes in Britain until 1830. However, the colony lagged in those areas that were deemed costly. For example, policing was given little priority on economic grounds. During the nineteenth century, a police force was developed for large cities, while rural policing was left in the hands of village headmen who were untrained, unpaid, and tended to be partisan towards their own relatives and friends. As such, the headmen largely acted in their own interests rather than as 'bureaucratic tools.'[26] The legal system, which during the early years of British rule consisted of a mixture of both British and native legal systems, was gradually reconstructed. In 1833, the Colebrooke-Cameron reforms reorganized the courts along utilitarian lines. A Supreme Court was set up to deal with serious crime and District Courts were responsible for judging lesser crimes within their geographical jurisdictions. In 1843 Police Courts were instituted to handle the minor crimes. The Police Courts were set up in part to handle the very large amount of litigation that took place amongst the locals. Rogers argues that the judicial system was a hybrid one,

> which did not coincide with either British or indigenous notions of justice but which was none the less compatible with local culture. Aided by court officials, lawyers and unofficial legal advisers, Sri Lankans used the courts as they used the spirits and demons of popular Buddhism. Both were perceived as amoral sources of power which responded more or less predictably to specific modes of address.[27]

Although the courts were popular with the islanders for allowing them access to the power of the state, in fact they 'generated little moral authority.'[28]

The principal British strategy for governing the various populations in Ceylon was twofold. First, elements of the pre-colonial feudal hierarchy were allowed to remain and keep many of their political functions. The British colonial government found these feudal lords useful in mediating between themselves and the Kandyan peasantry.[29] The nobles collaborated with the British in ways that ultimately destabilized the feudal land system. For example, the nobles found that it was advantageous to participate in the commercial land market, selling off their lands to outside interests without respect for the users' rights. The new landlords then often claimed rights to the services of tenant cultivators creating great hardship.[30] Second, British settlers and administrators were encouraged to gradually introduce British laws, institutions, manners, religion and morals into the country. Education in English language and culture was established in order to create a class of westernized elites and to spread Christianity. This was accomplished to a certain extent, as a western-educated Sinhalese elite found it to their advantage to act as agents of European theories of modernization, often espousing the same doubts as to

---

26 Ibid., 48.
27 Ibid., 9.
28 Ibid., 1.
29 Bandarage, *Colonialism*, 157.
30 Ibid.

the health, hygiene and discipline of their fellow Sinhalese and the Indian migrants as did the British.[31] They occupied what has been called 'political society' by Partha Chatterjee,[32] a mediating space between the people and the state. They took up the task of bridging the seemingly unbridgeable gap between the colonial administrators and the masses, who were looked upon as a population[33] of racialized bodies to be studied and controlled more than as a civil society in the classic normative European sense of the term.

London and Highland Ceylon: The Coffee Connections

Coffee had been used medicinally in Arabia since the tenth century, but did not become a popular drink there until the fifteenth. By the mid-seventeenth century increasing world trade had cultivated a taste among the bourgeoisie in Europe for orientalia and exotic luxuries, including coffee, tea, chocolate and tobacco.[34] Coffee owed much of its popularity in Europe to the fact that contaminated drinking water limited people to either fermented beverages or those made with boiled water such as coffee. By the mid-seventeenth century coffee was favoured over beer in many circles as it had acquired a reputation as the 'great soberer,' an anti-erotic medication and stimulant of the intellect.[35] This led to the development of a gendered geography of coffee houses as sites of enlightened public discourse reserved predominantly for men. Coffee houses in Europe were important as new public spaces where men could gather to discuss scientific and political ideas and conduct business.

The first coffee house in England opened in 1650, while the most famous was established by Edward Lloyd in 1688 as a meeting place for men in the maritime trades who came to hear trade news and buy insurance.[36] By the early years of the eighteenth century coffee houses in London outnumbered taverns overshadowing them in importance as places of public intercourse and important sites of knowledge transfer and decision-making. During the eighteenth century, insurance brokers rented booths in Lloyd's to transact business and by the end of the century it became known as Lloyd's of London, the insurance brokerage firm. The London stock exchange similarly emerged from Jonathan's Coffeehouse where a list of stock and

---

31 G. Prakash, 'Body Politic in Colonial India', in *Questions of Modernity*, edited by T. Mitchell (London: University of Minnesota Press, 2000), 205.

32 P. Chattergee, 'Two Poets and Death: On Civil and Political Society in the Non-Christian World', in *Questions of Modernity*, edited by T. Mitchell (Minneapolis: University of Minnesota Press, 2000), 42–3.

33 Chatterjee, ibid., 43, says of Foucault that he 'has been more perceptive than other social philosophers of recent times in noticing the crucial importance of the concept of population for the emergence of modern governmental technologies.' For Foucault's definition of population see M. Foucault, 'Governmentality', in *The Foucault Effect: Studies in Governmentality*, edited by G. Burchell, C. Gordon and P. Miller (Chicago: University of Chicago Press, 1991) 87–104; Chattergee, 'Two poets'; Legg, 'Foucault's Population Geography', and Prakash, 'Body Politic', 204.

34 W. Schivelbusch, *Tastes of Paradise* (New York: Vintage, 1993), 18–19.

35 Ibid., 34–35.

36 J. Thorn, *Le Café* (Koln: Taschen, 2001), 15, 49–50.

commodity prices was posted and organized trading in marketable securities began. Similarly, the East India Company based itself in the Jerusalem Coffeehouse, also in London, and the Royal Society, the most influential scientific institution of its day, was founded in a coffee house in Oxford. Thus on a variety of levels coffee houses were important nodes in Britain's global commercial network.

In Britain towards the end of the eighteenth century, coffee houses began to be eclipsed by men's clubs as social spaces in which to transact business. However, by this time, coffee had become a domestic drink not only for men, but for women who established ladies' coffee circles as a kind of anti-coffee house.[37] This greatly increased its consumer base. Even more important to the growth of consumption was the move away from mercantilism in the early nineteenth century, for until that time coffee had been taxed as a luxury good, dampening consumption. With increases in free trade, and the growth of an urban bourgeoisie, coffee consumption exploded by the mid-nineteenth century.[38] Overall, coffee consumption increased fifteen-fold during the nineteenth century, with Brazil by mid-century producing over half the world's coffee.[39] Although outstripped by tea, the consumption of coffee rose dramatically throughout the nineteenth century in Britain,[40] sustaining the networks which entwined the fate of planters, subsistence farmers, migrant labourers, merchants, and colonial administrators throughout Britain's tropical empire. However, as we shall see, non-human agents such as disease also circulated within these networks threatening their continuing stability and viability.

The Plantation Economy

Until the late seventeenth century, Europeans had purchased coffee from Arabia,[41] but as the European demand for coffee increased, the French, the Dutch and the British sought to produce coffee in their own colonies.[42] This move was fuelled by mercantilist fears of an outflow of capital and desire to increase state control over plantation production. Lowland coastal Ceylon had for many centuries served as

---

37 Schivelbusch *Tastes*, 69.

38 Topik and Clarence-Smith, 'Introduction', 6–7. The coffee consumption in Britain increased from one ounce per person per annum in 1801 to 1lb. 5½ oz. per person in 1841. I.H. Vanden Driesen, 'Coffee Cultivation in Ceylon (1)', *The Ceylon Historical Journal* 3 (1953), 41.

39 S. Topik, 'The Integration of the World Coffee Market', in *The Global Coffee Economy in Africa, Asia and Latin America, 1500–1989*, edited by W.G. Clarence-Smith and S. Topik. (Cambridge: Cambridge University Press, 2003), 31.

40 Whilst coffee became the mass-consumed drink of choice in America and much of northern Europe, Britain and Russia preferred tea and Spain, chocolate. Topik and Clarence-Smith, 'Introduction', 7.

41 M. Tuchscherer, 'Coffee in the Red Sea Area From the Sixteenth to the Nineteenth Century', in *The Global Coffee Economy in Africa, Asia and Latin America, 1500–1989*, edited by W.G. Clarence-Smith and S. Topik (Cambridge: Cambridge University Press, 2003).

42 There was a lucrative trade of re-exporting coffee to those consuming countries that did not have tropical colonies. Topik and Clarence-Smith 'Introduction', 6.

an important node in a trading network dominated first by Muslims, and after the 16th century, by Europeans (the Portuguese, the Dutch, and the British) linking East Asia and Europe.[43] After they captured the Maritime Provinces from the Portuguese in 1656, the Dutch convinced local peasants to expand their production of Arabica coffee and sell it to them at five stuivers per pound. By 1739 the annual coffee export from Ceylon had reached 100,000 pounds. However, in 1690 the Dutch introduced coffee into their colony of Java where they could produce it more cheaply. They then began to actively discourage its production in Ceylon by refusing to pay more than two stuivers per pound. They did so because they were afraid that the Ceylonese production would lower the value of Dutch coffee on the European market. So successful was this Dutch policy that Ceylonese coffee production had dropped four fifths by the end of the eighteenth century.[44]

When the British captured the Maritime Provinces from the Dutch in 1796 they sought to revive peasant coffee cultivation more than doubling peasant production from 94,500 pounds in 1806 to 216,500 in 1813.[45] However, the finest coffee-growing region in the highlands remained under the control of the Kandyan king until 1815. Almost immediately after the British took control of the Kandyan highlands, they began to grow coffee for export. By the 1840s a plantation economy had been established and coffee had become the primary raison d'etre of Ceylon as a British crown colony. The region was increasingly exposed to distant events as its agriculture was restructured and inserted into a vast colonial network as an export-oriented region. Early British mercantilist policies, which concentrated on trade monopolies and tax revenue rather than production, had discouraged large-scale European or native capital investments outside of Colombo. But by the 1830s such policies were increasingly challenged by competing theories of free enterprise and new political rationalities of colonialism.

---

43 It appears that the Arabica beans that were planted were brought to India by Muslim pilgrims in the seventeenth century and that later in the century the transfer of coffee to Java was from Indian seedlings again imported through Muslim traders. On this see W.G. Clarence-Smith, 'The spread of coffee cultivation in Asia from the seventeenth to the early nineteenth century', in *Le Commerce du Café Avant L'etre des Plantations Coloniales*, edited by M. Tuchscherer (Cairo: Institut Français d'Archéologie Orientale, 2001) 371–84.

44 Vanden Driesen, 'Coffee (1)', 31–32. Although the Dutch did not use slave labour, it was heavily coerced in West Java, where they had contracts with local rulers who forced their subjects to collect a quota. Each village had to send the produce of several hundred trees. Coffee was subsequently spread by the Dutch and the French from Java to the Americas, via their home botanic gardens. So successful was this transfer to the Americas that by the late eighteenth century, 80 per cent of world coffee production was in the new world. (Topik, 'The integration,' 28–29.) On the Dutch experience see: M.R. Fernando, and W. O'Malley, 'Peasants and Coffee Cultivation in Cirebon Residency', in *Indonesian Economic History in the Dutch Colonial Era*, edited by A. Booth et al. (New Haven: Yale University Press, 1990) 171–86; and on the Portuguese in Asia see W.G. Clarence-Smith, 'Planters and Small-Holders in Portuguese Timor in the Nineteenth and Twentieth Centuries', *Indonesia Circle* 57 (1992): 15–30.

45 Vanden Driesen, 'Coffee (1)', 32–33.

Plantations in the British colonies became sites of the adaptation and further development of industrial methods and other western technologies of rationalization, calculation, and discipline.[46] In fact, it has often been claimed that in the mid-to-late nineteenth century, Europe's tropical colonies had become laboratories of modern governmentality.[47] Kurian argues that there were three principle characteristics of labour regimes in industrial plantations: the significant role of migrant labour, extreme control over labour, and an interventionist role of the colonial state in the recruitment and control of labour practices.[48] A possible fourth characteristic was the role of a growing international knowledge economy which the Ceylon government and planting community was able to tap into (and contribute to) in forwarding the interests of the coffee industry. Mokyr identifies a 'knowledge revolution' that had its roots in the late eighteenth century – a vast assemblage of 'prescriptive knowledge' as he terms it.[49] Thrift says that this knowledge revolution depends upon what he calls the development of a whole 'enterprise ecosystem' composed of 'makeshift institutional responses' on the part of states, non-state institutions and individual entrepreneurs all functioning to disseminate knowledge relevant to furthering the economy.[50] Given the general lack of funding for the Ceylon government and the state of world communication at the time, 'makeshift' is a very apt term. The relative newness and contested nature of such active governmentality lent a decidedly ad hoc tone to joint efforts at producing prescriptive knowledge about planting. Nevertheless, as we will see, such knowledge production became an important aspect of colonial governmentality in Ceylon.

Such modern plantations were established nearly simultaneously in British South Asia, the Dutch East Indies, French Indochina, and the Philippines. As Kurian[51] points out, there was an intense international competition among industrial plantations, with each having its own somewhat different trajectory. Consequently producers had little control over market prices which were established in Western Europe. Thus places like Ceylon were not self-contained, self-sufficient nodes in a global network; they were both active and reactive as participants in various coeval, but entangled

---

46 P.P. Courtenay, *Plantation Agriculture* (London: Bell and Hyman, 1980), 55–56; R. Kurian, *State, Capital and Labour in the Plantation Industry in Sri Lanka 1834–1984* (Amsterdam: University of Amsterdam, 1989), 1, 10; H. Gregor, 'The Changing Plantation', *Annals of the Association of American Geographers*, 55 (1965): 221–238.

47 Metcalf, *Ideologies*, 29, argues that India was a laboratory for the creation of the liberal administrative state in the mid-nineteenth century. For a discussion of Dutch colonialism as an experiment in modernity see, Stoler, *Race*, 15. For the French case see, P. Rabinow, *French Modern: Norms and Forms of the Social Environment* (Chicago: University of Chicago Press, 1989).

48 Kurian, *State*, 10–12.

49 J. Mokyr, *The Gift of Athena: Historical Origins of the Knowledge Economy* (Princeton: Princeton University Press, 2001).

50 N. Thrift, *Knowing Capitalism* (London: Sage, 2005), 95. Thrift's main concern is with a later phase of this knowledge revolution, which he terms the cultural circuit of capitalism. However, his description of a makeshift enterprise ecosystem seems to apply well to this early period of governmentality in Ceylon.

51 Kurian, *State*, 9–10.

projects of modern capitalism. Although plantation agriculture required much more capital investment than transit trade, it was believed that by controlling plantation agriculture, the British could gain more effective control over the volume of that trade. Plantations were now expected to produce a net profit for the government. The governors therefore encouraged the establishment of export agriculture.

Although the first European commercial plantations in the highlands were created in the 1820s, the British colonial officials who set these up generally lacked experience and most eventually lost them,[52] most often to other more experienced British planters.[53] In the meantime they had used their official positions to set up an infrastructure favourable to the establishment of large-scale coffee planting. The removal of export duties on coffee and taxes on coffee lands and the building of transportation networks during the rule of Governor Barnes were examples of state intervention on behalf of the plantation economy.[54] Also significant was the relocation of the government botanical research station from the capital in Colombo to Peradenya near the plantations where scientific research on the growing of coffee plants and their diseases was carried out.[55] Shifts in transportation technology such as the development of roads and later the railway were central in the shift away from early forms of exploitation of peasant grown crops to plantation production that depended upon the efficient movement of large numbers of people and goods.[56] In 1841 The Ceylon Bank was opened to finance the expanding coffee plantation enterprise.

It is not surprising that for such a large-scale transformation to be accomplished in the highlands, related events of great moment would have to have happened elsewhere. The first was the abolition of slavery in the West Indies in 1833; this caused plantations there to collapse and severely disrupted the supply of West Indian coffee to Europe.[57] By the end of that decade, British coffee production had declined

---

52 See Vanden Driesen, ' Coffee (1)'; I.H. Vanden Driesen, 'Coffee Cultivation in Ceylon (2)', *The Ceylon Historical Journal* 3 (1953): 156–72; A.C.M. Ameer Ali, 'Peasant Coffee in Ceylon During the Nineteenth Century', *Ceylon Journal of History and Social Science* 2 (1972), 50; D. Moldrich, *Bitter Berry Bondage* (Colombo: Ceylon Printers, 1989); E. Meyer, 'Enclave Plantations, Hemmed-in Villages and Dualistic Representations in Colonial Ceylon', *Journal of Peasant Studies* 19 (1992): 199–228.

53 After the 1850s the colonial government embarked on a systematic effort to recruit new planters with previous experience in managing farms or plantations (D. Forrest, 'Hundred Years of Achievement', *The Times of Ceylon Tea Centenary Supplement*, 31 July 1969.)

54 See, G.C. Mendis (ed.), 'Concessions for the Cultivation of Certain Agricultural Products, Regulation Number 4 of 21 September 1829', in *The Colebrooke Cameron Papers. Documents on British Colonial Policy in Ceylon 1796–1833*. Volume 2, edited by G.C. Mendis (Oxford: Oxford University Press, 1956), 279.

55 L.A. Mills, *Ceylon under British Rule, 1795–1932* (London: Oxford University Press, 1933), 224; Kurian, *State*, 47; On the role of science in the early plantations see, T.J. Barron, 'Science and the Nineteenth Century Ceylon Coffee Planters', *The Journal of Imperial and Commonwealth History* 16 (1987): 5–21.

56 Kurian, *State*, 9; Perera, *Society and Space*, Ch. 3.

57 See S. Smith, 'Sugar's Poor Relation: Coffee Planting in the British West Indies, 1720–1833', *Slavery and Abolition* 19 (1998): 151–72. While the British ended slavery in 1833, the French did not end it in their territories until 1848, the Americans and Dutch in

substantially; in Jamaica alone 465 coffee plantations totalling more than 188,000 acres were abandoned.[58] The second was the passage of two key pieces of legislation. One was the equalization of the duty on West Indian and Ceylonese coffee in 1835, (coffee from South Asia had previously been charged a 50 per cent higher tariff) and the other was the Crown Lands Encroachment Ordinance of 1840, which seized for the British Crown all land that had not been continuously under cultivation and for which no deed or tax receipt could be shown.[59] These were key events in the creation of a boom in the highland plantations attracting farmers from Britain and planters from other British colonies. These were heady times for as one commentator said, 'if the progress of [European] cultivation continues to advance at the same rate that it has done for the last five years, an immense alteration will be effected in the heretofore desert wastes of the island, and, as a necessary consequence, in the moral character and intellectual advancement of its inhabitants.'[60] As Perera points out, referring to the chena lands of peasants as unoccupied or uncultivated waste land was a way of linguistically clearing land that European capitalists wished to claim for themselves.[61]

The Crown Lands Encroachment Ordinance was highly detrimental to the Kandyan peasantry in certain areas where few had satisfactory documentation.[62] They lost much of the common land they had used for (chena) cultivation and grazing, a vital supplement to their rice lands. They did not however, lose their rice lands in the wet valley bottoms as these were not good for coffee cultivation. In 1833 before this legislation, only 146 acres of crown lands had been sold, but in 1840 when a buying frenzy took hold, 78,686 acres were sold, 13,275 acres in one day alone. Most land was sold to British officials at the knock-down price of five shillings an acre.[63] Between 1840–43, over 230,000 acres of crown lands were sold, mostly for coffee estates.[64] Although in the early years a sizable percentage of coffee lands

---

1863, the Spanish in 1885 and the Brazilians in 1888. During the nineteenth century, England imported coffee from the West Indies, Java, Ceylon, and Brazil.

58 D. Hall, *Free Jamaica, 1838–1865: An Economic History* (New Haven: Yale University Press, 1959), 184, cited in Webb, *Tropical*, 188.

59 For a study of a similar confiscation in French Indochina see, M. Cleary, 'Land Codes and the State in French Cochinchina, c. 1900–1940.' *Journal of Historical Geography* 29 (2003): 356–75.

60 L. De Butts, *Rambles in Ceylon* (London: W.H. Allen, 1841), 185.

61 Perera, *Society and Space*, 64.

62 Vanden Driesen, I.H. 'Land Sales Policy and Some Aspects of the Problem of Tenure, 1836–86. Part 2', *University of Ceylon Review* 15 (1957), 40. There has been a debate about the impact of the Ordinance on peasant livelihood. Jayawardene has argued that the coffee estates after the first few years had no interest in chena lands because they were unproductive for coffee. Consequently, he argues, the estates had a very limited impact on peasant agriculture. See, L.R. Jayawardene, *The Supply of Sinhalese Labour to Ceylon Plantations (1830–1930): A Study of Imperial Policy in a Peasant Society* (Ph.D. Thesis, University of Cambridge, 1963).

63 Lewis, *Coffee*.

64 D. Snodgrass, *Ceylon: An Export Economy in Transition* (Homewood, Illinois: Richard Irwin, 1966), 22.

were in the hands of native farmers, much of this was in small holdings. There were only a few Sinhalese and Tamil plantation owners in these years.[65] Their numbers were limited in part by the British-owned banks and agency houses' policies that either refused to loan them sufficient funds[66] or charged them higher rates than were charged to British entrepreneurs.[67] However, peasant coffee grown on small plots remained important throughout the coffee period. During the 1820s and 1830s, before the estate boom, smallholder coffee increased ten-fold.

Table 2.1   Average exports per year of native coffee, 1808–1833[68]

1808–1813	261,500 lbs
1816–1820	434,800 lbs
1821–1825	1,100,000 lbs
1826–1833	2,061,400 lbs

Through the plantation boom of the 1850s and 1860s, smallholder coffee still accounted for nearly a quarter of total output and at its peak reached 50,000 acres. During these peak years, many plots were small and nearly 100,000 villagers were involved in coffee growing.[69] Whilst crown lands were purchased mainly by Europeans during the early decades, after 1860, 68 per cent of the crown lands grants were purchased by peasants. However, there continued to be a marked difference in size of purchase by race. For example, 77 per cent of peasant purchases were of less than five acres and 37 per cent less than an acre. European purchases during this period, on the other hand, were all over ten acres and 25 per cent were over 50 acres.[70] Having said this, there were some very large estates that were owned by non-Europeans. For example, in the early 1870s the Sinhalese capitalists Jeronis Pieris and Charles Soysa had respectively 1,000 and 2,400 acres in coffee, while the Moor brothers Sinne Lebbe in the mid-1860s had 2,300 acres under crop.[71]

---

65 Meyer, 'Enclave', 207.

66 It cost a minimum of 3,000 pounds to establish the typical estate in Ceylon during the 1840s (Snodgrass, *Ceylon*).

67 M. Roberts and Wickremeratne, L.A. 'Export Agriculture in the Nineteenth Century', in *University of Ceylon, History of Ceylon,* Vol. 3, edited by K.M. de Silva (Colombo: University of Ceylon, 1973), 94–97.

68 Vanden Driesen, 'Coffee (1)', 35.

69 Snodgrass, *Ceylon*, 29.

70 K. Jayawardena, *Nobodies to Somebodies: The Rise of the Colonial Bourgeoisie in Sri Lanka* (Colombo: The Social Scientists' Association, 2000), 141.

71 Ibid., 176, 178, 221.

Table 2.2　The Ceylon coffee industry 1834–86[72]

Year	Export volume (000 cwts)	Export unit value (s./cwt)	Area planted (000 acres)	Yield (cwts/ acre)
1834	26	30	–	–
1835–39	46	47	–	–
1840–44	97	49	23	4.2
1845–49	260	33	51	5.1
1850–54	344	40	59	5.9
1855–59	537	48	138	3.9
1860–64	615	51	199	3.1
1865–69	939	52	243	3.9
1870–74	881	66	276	3.2
1875–79	795	108	310	2.6
1880–84	433	89	259	1.7
1885	316	78	139	2.3
1886	179	89	110	1.6

Production skyrocketed during the early 1840s and then crashed in 1847. This was brought on by the lowering of the duty on Brazilian and Javanese coffee as part of Peel's policy of greater free trade, thereby precipitating a drop in price on the British market. This was followed by a general depression in Europe from 1847–49 and a lowering of demand for coffee and drying up of credit for estates. In consequence, the price for coffee plummeted from 41 shillings per hundredweight in 1845, to 28 shillings in 1848.[73] Consequently, ten per cent of the estates being farmed in 1847 were abandoned by 1849.[74] Estates were sold at this time for a fraction of what they had been worth a few years before.[75] During the 1850s the area under coffee tripled and grew at the rate of 35,000 acres per year during the boom of the 1860s. Given that for various reasons, plantations were operating much more efficiently than during the 1840s and because the price of coffee had stabilised in London during that period at around 54 shillings per hundred weight, profits were still good.[76] Between 1850 and 1870 coffee production again prospered. During this period, companies replaced many private owners, but even then, two thirds of estates remained in individual hands and one third were farmed by owner-operators while the other third were managed for individuals by agency houses.[77] By 1857 new technologies allowed an acre of forest land to be brought under cultivation for one tenth of its cost in

72　Snodgrass, *Ceylon*, 20.
73　Vanden Driesen, 'Coffee (1)', 49–53.
74　Ibid., 55.
75　Snodgrass, *Ceylon*, 18.
76　Vanden Driesen, 'Coffee (2)', 160.
77　Snodgrass, *Ceylon*, 27

1844.[78] Ten years later, another piece of technology, the Colombo-Kandy railroad cut transportation costs from the estates by between 60 and 75 per cent.[79] Before the railway, the journey by cart took eight to ten days and coffee couldn't be shipped during the height of the monsoon because of rain-damage to the beans and because the roads often became impassable.[80] The maintenance of roads was a major drain on the revenue of the colony. The combination of heat, torrential rains and the thin iron-rimmed wheels of carts loaded with as much as 45 hundred weight of coffee could destroy the surface of a road in as little as two months. In fact, by the late 1840s the maintenance of roads in the coffee districts was consuming 15 per cent of the island's revenues.[81]

In 1869, the coffee leaf rust disease *Hemileia vastatrix* struck with great force and progressively spread over the next five years eventually affecting all the estates on the island. This coffee rust disease spread rapidly as large estates provided open spaces where once forests predominated. There were now few barriers to prevent the spread of fungal spores on the winds. As is often the case with drastically changed ecologies, new agencies appear and become powerful. As we shall see, this increasingly open environment also provided excellent breeding areas for mosquitoes and it is likely that malaria, which was introduced from the mainland at an early date, was exacerbated by the pattern of ecological change.[82]

While the coffee rust disease greatly reduced yields, many producers were initially saved by a dramatic rise in the price of coffee on the world market. The price of coffee shot up by nearly 59 per cent in a few years in the early 1870s. This rise in price was brought on by increased demand in the American market coupled with short crops in Ceylon's two main competitors, Java and Brazil.[83] This more than sufficed to compensate planters for declining yields due to spreading leaf disease.[84] Clarence-Smith[85] points out that as the century progressed there was increasing volatility in world coffee prices. Prices on the international market rose steadily from the 1840s until they peaked in the mid 1870s. After 1870, control over the coffee market swung increasingly away from producers into the hands of importers. The key reason for this shift was the laying of a transatlantic telegraphic cable connecting South America to coffee exchanges in New York and London. From 1870 onwards, importers instantly learned the supply levels and prices in the Latin

---

78 Bandarage, *Colonialism*, 77

79 Snodgrass, *Ceylon*, 31. On the politics of the building of the Colombo-Kandy railway, see I. Munasinghe, 'The Colombo-Kandy railway', *The Ceylon Historical Journal* 25 (1978): 239–49.

80 I.H. Vanden Driesen, *Indian Plantation Labour in Sri Lanka: Aspects of the History of Immigration in the 19th Century* (Nedlands: University of Western Australia, 1982), 76.

81 I.H. Vanden Driesen, 'Some Trends in the Economic History of Ceylon in the "Modern" Period.' *Ceylon Journal of Historical and Social Studies* 3 (1960), 9–10.

82 Webb, *Tropical*, 10–12.

83 Webb, *Tropical*, 110.

84 Vanden Driesen, 'Coffee (2)', 163.

85 W.G. Clarence-Smith, 'The Coffee Crisis in Asia, Africa, and the Pacific, 1870–1914', in *The Global Coffee Economy in Africa, Asia and Latin America, 1500–1980*, edited by W.G. Clarence-Smith and S. Topik (Cambridge: Cambridge University Press, 2003), 101.

American and Asian producing regions. As a result international grading and pricing became increasingly uniform.[86] During the 1870s, although much of the good coffee land had already been sold, land sales continued briskly as speculators invested due to the high prices for coffee on the world market.[87] In fact the very high cost of new coffee lands in the 1870s was part of the reason that so many plantations failed.[88] By the early 1880s the good times were over for Ceylon coffee planting as it was subjected to a devastating conjuncture of events: the Great Depression in England from 1879 to 1884 (almost all of Ceylon's coffee went to the British market with only a tiny fraction re-exported[89]), strong competition from lower-grade Brazilian coffee, and most notably the spread of the coffee fungus disease.[90]

Throughout the 1870s and increasingly during the early 1880s as coffee failed, planters experimented with alternative plantation crops. Cinchona for the production of quinine was grown throughout the 1870s and 1880s, but by the mid 1880s, so much had been produced that the bottom dropped out of the world market and production was virtually abandoned.[91] Cocoa was tried at lower altitudes, but never became a viable alternative to coffee. Tea planting, on the other hand, began slowly in the 1850s and grew by leaps and bounds throughout the early 1880s as it became increasingly clear that coffee would never recover. By 1886, it had surpassed coffee in terms of acreage. While many planters failed to make the transition, those whose higher altitude plantations were suitable for tea, found that they had many of the ingredients to make the switch: idle land, a ready workforce, and an organizational system that could be adapted to tea. The villagers whose livelihoods were wiped out by the coffee disease were, as we shall see, less able to make this transition in large part because tea was a more capital-intensive crop that lent itself less well to small-scale cultivation.[92]

86 Topik, 'The integration', 39.

87 Snodgrass, *Ceylon*, 23.

88 Mills, *Ceylon*, 246.

89 A.M. Ferguson, and J. Ferguson *The Ceylon Directory for 1874* (Colombo: A.M. and J. Ferguson, 1874), 243.

90 Vanden Driesen, 'Coffee (2)', 165.

91 Ibid., 170.

92 Snodgrass, *Ceylon*, 32.

Chapter 3

Dark Thoughts: Reproducing Whiteness in the Tropics

'Europeanness' was not a fixed attribute, but one shaped by environment, class contingent, and not secured by birth.

A. Stoler[1]

the government of oneself, that ritualization of the problem of personal conduct.

Michel Foucault[2]

Introduction

From the 1830s until the 1880s, young British men went to Ceylon to make their fortunes as coffee planters. The typical planter of the period was the second son of a family of moderate means,[3] although, as we shall see, there was a difference in class between those who bought coffee estates and those who went out as managers and also between the first planters and a second generation who came out during the boom years in the 1870s.[4] British women did not to follow the planters into the hills in significant numbers until the end of the period. For many young men their house on the plantation was a first home of their own where it was intended that the discipline of the family would be replaced by manly, Christian self-discipline.[5]

---

1   Stoler, *Race*, 104.

2   Foucault, 'Governmentality,' 87.

3   Kurian, *State*, 26. An anonymous account written in the mid-1880s described the planters who came out during the boom years of the early 1870s as follows: 'They were mostly of the same class as my husband – young fellows who had had a good public school education, and who had gone in for examinations, to pass for the Army or Civil Services, and had failed. Many of them were University men who had taken an ordinary degree and finding nothing but the scholastic professions or Holy Orders before them had come out to Ceylon in the hope of finding more congenial work.' Anonymous, *Fickle Fortune in Ceylon* (Madras: Addison and Company, 1887), 36. Quoted in Barron, 'Science,' 17.

4   The later planters were of a different class from many of the managers who were recruited during the boom of the 1840s. The frenzy of speculation in coffee planting in the 1840s was such that many working class men were recruited directly off ships and the army. Planters later in the century tended to be from wealthier families than the early planters. They were often looked upon as inexperienced and not as manly or heroic by the ones who first opened the estates.

5   On the history of evangelical, Christian manliness and the enthusiasm for the ideal amongst early Victorian middle classes, see J. Rutherford, *Forever England: Reflections on Masculinity and Empire* (London: Lawrence and Wishart, 1997). Manliness included physical

Within the colonies as in Britain, the concept of self-discipline was racialized, classed, and gendered. In the colonies self-discipline was seen as productive of racial and class hierarchy and as a crucial foundation for the reproduction of a ruling class. Controlled self-conduct was seen as necessary for the planter not only to signal his authority, but also to present himself as a proper role model for the native elite and ideally for the native population as a whole.

The nineteenth century middle class home in England was considered the place where bourgeois values and consciences would be cultivated. It was also the principal site where sexuality was regulated, where impulsive passions could be tamed and controlled.[6] As we will see, this spatialization of sexuality presented difficulties for British homes in the colonies, especially as they were commonly occupied by single men. Thus, one could argue that at another scale, the spatialization of sexuality produced 'shadowy regions' in the city,[7] and in the case of the colonies, shadowy regions of the world, which allowed for a degree of tolerance and relaxation of middle class expectations.[8] Despite bourgeois ambitions, in practice Ceylon was seen as a frontier, a place where metropolitan norms of respectability were in fact relaxed, at least to a degree, while still remaining the primary mark of whiteness.[9] The resulting tension within colonial homes produced ambivalence about the standards of appropriate behaviour. This tension is evident in the self-censuring of written accounts of planter life. Although these tended to be silent about relations with women, occasionally young planters expressed acute feelings of loss of self-control in their letters home. In fact, the planters were often painfully aware that they did not live up to high bourgeois standards and expressed fear of discrediting carefully cultivated British notions of racial superiority. This loss of control was equated with a loss of Europeaness, leading to anxiety about racial, class, and gendered identities and differences.

---

courage, a taste for adventure, as well as being a gentleman. Self-discipline was one of the most important values of evangelicalism in Victorian England. In fact Evangelicals saw the lack of moral restraint as the problem of the age. On this in relation to the home, see C. Hall, *White, Male and Middle Class: Explorations in Feminism and History* (Cambridge: Polity Press, 1992); J. Tosh, *Manliness and Masculinities in Nineteenth-Century Britain: Essays on Gender, Family and Empire* (London: Pearson Longman, 2005), and Wiener, *Reconstructing the Criminal*.

6   Foucault, *History of Sexuality Vol. 1*.

7   T. Osborne and N. Rose, 'Governing Cities: Notes on the Spatialization of Virtue', *Environment and Planning D: Society and Space* 17 (1999): 737–60.

8   P. Howell, 'Foucault, Sexuality, Geography', in *Space, Knowledge and Power: Foucault and Geography*, edited by J. Crampton and S. Elden (Aldershot, Hampshire, 2007) 291–316.

9   Respectability was of great importance to colonials. Lambert has shown that poor whites were not considered respectable and thus were unsettling figures for middle class whites in the tropics. Lambert, *White Creole*, 81. Poor white women in particular were unwelcome in the colonies because it was feared that they would consort with natives or fall into prostitution discrediting the Europeans. On this see P. Levine, *Prostitution, Race and Politics: Policing Venerial Disease in the British Empire* (New York: Routledge, 2003), 234–35.

As I will show, the British planters were often keenly aware that their behaviour was judged negatively by the local population and the other British living in Ceylon at the time. Lord Torrington wrote of the planters to Earl Grey,

> the mass of the coffee planters, many of them the very worst class of Englishmen has very much tended to lower and degrade our caste and character in the eyes of the natives. The habits of the planters, particularly with the women, has been most objectionable and in the eyes of a Kandyan a coffee planter is a term of reproach.[10]

The issue of the sexuality of Europeans was not only about maintaining social distance and propriety, but had a strong bio-political component as well.

As Ann Stoler argues, the self-disciplining of individual colonial male Europeans was seen as 'tied to the survival of all Europeans in the tropics and hence the bio-politics of racial rule.'[11] The Europeans thought of themselves not simply as individuals, but as members of a racial population whose very survival was endangered. The concept of self-discipline was deeply intertwined with bio-politics.[12] Howell uses the term 'biopolitics of sexuality' to refer to the idea that Victorian discourses and practices of sexuality were not just about discipline, Puritanism, repression and bourgeois hypocrisy, but about spatializing sexuality, the creation of places of tolerance and freedom, while keeping the bio-political security of the colony firmly in mind.[13] This was a difficult balancing act that often failed both politically in terms of low levels of cultural hegemony and psychologically for individual planters who had internalised British fears of racial degeneration. While the colonies were alternatively seen as laboratories of modernity, as heterotopic, marginal, or individualistic spaces of tolerance, the relations of British men with native women were seen as undermining the goal of reproducing a biologically strong and legitimate ruling elite.

---

10 Torrington to Grey, 11 August 1848, quoted in K.M. De Silva (ed.) *Letters on Ceylon 1846–50. The Administration of Viscount Torrington and the Rebellion of 1848* (Kandy: K.G.V. De Silva and Sons, 1965), 98.

11 Stoler, *Race*, 45.

12 Agnew and Coleman take the Foucauldian position that it is necessary to look at bio-politics both at the state level and beyond at a geo-sociology of power. The idea is to look beyond top down discipline and control of docile bodies to the micro-practices that divide, isolate, and fragment power – subcontracting it to remote and specialized authorities. They also argue in line with Foucault that there are technologies of government based on a bio-political model of security and welfare that operate on self-scrutiny as much as through the regulation of a population. Self-discipline is an important component of bio-politics. J. Agnew, and M. Coleman, 'The Problem With Empire', in *Space, Knowledge and Power: Foucault and Geography*, edited by J. Crampton and S. Elden (Aldershot, Hampshire, 2007) 317–39.

13 Howell, 'Foucault', 305.

Moral Hygiene and the Tropical Home

Dreams of an economic and cultural transformation of tropical nature by British capitalism were pervasive in mid-nineteenth century Ceylon as elsewhere.[14] For example, the planter William Boyd[15] wrote:

> [a] new era is dawning on Ceylon... The steam engine will be heard in every hollow, the steam horse will course every valley; English homes will crown every hillock, and English civilisation will bless and enrich the whole country, causing the wilderness to blossom as a rose, and making Ceylon, as it was in former times, a garden of the world and the granary of India.

It is important to note that in the imagination, the tropics do not become merely an extension of Britain, but are transformed into proper colonies through Britain. They remained subservient to the interests of Britain, but nevertheless are thoroughly transformed through British capitalism, British civilisation and British domestic values. However, in the hills of mid-nineteenth century Ceylon such a sweeping transformation remained a dream. In fact, the combined impact of the tropical climate and native culture was seen as possibly even more transformative of the British than vice versa. The well-known medical authority Henry Marshall had the following to say about the impact of Ceylon's climate on newly arrived Europeans,[16]

> The inhabitants of the high latitudes, on their arrival in Ceylon, for the most part, undergo some change, both in their physical and mental functions. Their constitutions become irritable, and easily affected by stimuli. Many of them soon sustain a deviation, more or less, from sound health, accompanied by a certain degree of emaciation. The skin loses the ruddy hue of robust health, and assumes a pale yellowish shade: moderate exercise becomes fatiguing, and the mind indisposed to much application. This observation is, however, liable to a number of exceptions: individuals of the higher classes, who endure little fatigue, live temperately, and do not expose themselves much to the direct rays of the sun, are not greatly liable to attacks of acute disease.[17] No care, however, can entirely

---

14 For a discussion of this trope of the European conquest of the tropics, see M.L. Pratt, *Imperial Eyes: Travel Writing and Transculturation* (London: Routledge, 1992); and D. Spurr, *The Rhetoric of Empire: Colonial Discourse in Journalism, Travel Writing and Imperial Administration* (Durham: Duke University, 1993).

15 W. Boyd, 'Autobiography of a Periya Durai', *Ceylon Literary Register* 3 (1888), 410.

16 H. Marshall, *Notes on the Medical Topography of the Interior of Ceylon* (London: Burgess and Hill, 1821), 74–75.

17 It is difficult to interpret medical opinion from a very different perspective, but it is safe to assume that Marshall was actually observing rather than merely imagining changes to European bodies. Perhaps his reference to yellow skin refers to jaundice, as liver disease was common in Ceylon. Or perhaps it was in part a reference to what we would now call a light tan. Tanned skin was not thought to be attractive amongst the bourgeoisie until the 1930s (on this see M. Blume, *Côte d'Azur: Inventing the French Riviera* (London: Thames and Hudson, 1992), 74.) Perhaps his discussion of the impact of class on health is a correct observation based on the fact that those classes were less likely to get tanned or suffer from certain diseases associated with the countryside.

prevent the debilitating influence of the climate, and a tendency to chronic disease of some of the vital organs. Senescence frequently precedes old age at a considerable distance.

While it was assumed that no one could permanently escape the impact of the tropics, one could buy time through proper bodily practices; time enough perhaps to make one's career and return home to Britain. The belief in the inevitable degeneration of Europeans in the tropics destabilized notions of progress, development and the civilising mission, contributing to a crisis of British, masculine identity.[18] I wish to make it clear from the outset that the ambivalent transcultural impacts, insecurities and anxieties found on the margins of the British empire should not in any way be conflated with a fragility in the political economy of empire. The British rule over this part of empire was secure during this period and consequently our understanding of its exploitative impact should not be diminished through attention to British psyches.[19]

While planters worried about racial degeneration, of even greater concern was disease, which often struck with sudden and catastrophic consequences. Shortly after his arrival the young planter, James Taylor[20] became seriously ill with dysentery and thought he was going to die. The planter Millie[21] wrote that most newly arrived planters came down with a 'seasoning fever' and if they survived this initial test of endurance, they would become acclimatized.[22] Unfortunately, this was often not the case. Taylor's friend David Moir, for example, died suddenly of cholera five years after his arrival. Taylor[23] writes how Moir took ill at eight in the morning and was dead by four in the afternoon.

In the tropical environment, the British found death lurking everywhere, even in natural forms that appeared most familiar. On September 18th, 1866 newly-arrived planter L.W. Daniell[24] wrote sadly to his mother in England:

> On Thursday morning a Mrs Corbet brought ... a bundle of vegetables amongst which was something that looked exactly like young rhubarb and was supposed to be meant for a hash or pudding. This was made for our dinner and was to have been sent in cold, but my Emmie after a hearty dinner of meat and curry asked for pudding and this being the only sweet ready, it was brought in hot. She ate nearly a plateful and my Georgie only about

---

18 S. Gikandi, *Maps of Englishness: Writing Identity in the Culture of Colonialism* (New York: Columbia University Press, 1996); R. Williams, *Culture and Society, 1780–1950* (Harmondsworth: Penguin, 1963); R. Phillips, *Mapping Men and Empire: A Geography of Adventure,* (London: Routledge, 1997), 13.

19 On this point, see B. Parry, *Delusions and Discoveries* (London: Verso, 1998).

20 J. Taylor, *Papers of James Taylor (1835–92).* (MSS. 15908-10). National Library of Scotland. Department of Manuscripts, 1851–92, 23 June 1852.

21 Millie, *Thirty.*

22 The terms seasoning and acclimatization were synonymous (Livingstone, 'Tropical climate', 101). The idea of seasoning remained in regard to some diseases while it fell away as a belief about the possibility of whites ever acclimatizing fully to the tropics. It remained an unsettled issue throughout the nineteenth century. Curtin, *Death By Migration,* 47.

23 Taylor, *Papers,* September 17, 1857.

24 L.W. Daniell, 18 September 1866. *Wentworth-Reeve Papers* (Centre for South Asian Studies, Cambridge University, 1866).

two teaspoonfuls when Fanny [the nurse] saw by her face she did not like it – she (F) then tasted it and said: "Oh! how nasty! How can you eat it Miss Emma?" Emmie then said she did not like it much and did not eat any more.

That night the children became violently ill and by morning he could feel little pulse on his daughter Emmie.

> She was in nurse's arms and at 9:30 her head was raised to try and get a little brandy down her throat and I heard it rattle, so I laid her head down again and my darling angel child whom I loved as perhaps nothing of the earth should be loved, fell asleep in Jesus. I thank God we were all able even in that moment to say 'His will be done.' Then I turned to Georgie who had the same symptoms. She fell asleep even more quietly than my Emma. Not a struggle of any kind, but a slight gasping for breath, and then slower and slower till at last a deeper one came and no more. I felt turned to stone, and as if no further sorrow could afflict me. I carried my bright blue-eyed caressing darling to be laid beside the sister she loved so well and a more peaceful beautiful sight mortal eyes never beheld than those two together.

Potential danger was sensed everywhere; in miasmas that produced fevers, in poisonous snakes, and in a seemingly innocent plant that appeared to be rhubarb. Although recently arrived Europeans didn't know much about tropical plant life, they sometimes found themselves trusting familiar appearances rather than checking with the Sinhalese. And it was not only Europeans who died from eating unknown plants; immigrant labourers from India frequently died from eating poisonous yams that they collected in the jungle unaware that the Sinhalese used them to commit suicide.[25]

Devastating to the British and Kandyans alike was the cholera epidemic of 1845 where one fifth of the population of Kandy died within a few weeks.[26] Boyd looking back on the early days of planting wrote,

> So great has been the mortality amongst young men whom I knew, that out of 17 who lived 27 years ago within a few miles of me in one district, only two, myself and another, now survive; and this I fear is also the experience of the few survivors of the old days of which I write.[27]

Extreme feelings of vulnerability produced a sense of embodiedness that the British had normally associated much more with women, the lower classes and racial others, and this was deeply disturbing to British men.

As there were far more British men than women in the hills, diaries, reminiscences, and recruitment literatures were scripted as narratives of heroic masculinity. The ideal was challenged, however, by the material realities of the environment including often utter failure to succeed in business or even to physically survive.[28] While earlier models of moral masculinity were explicitly evangelical, later constructions

---

25 The *Ceylon Observer*, January 22, 1859.

26 W. Boyd, 'Ceylon and its Pioneers', *Ceylon Literary Register* 2 (1888), 281.

27 Ibid., 275.

28 On colonial failure, see Thomas and Eves, *Bad Colonists*.

became more secularised into a cult of character, the gentleman. As Rose points out, the Victorian notion of character was widely dispersed across many different practices, including not only heroic pioneering adventures, but also in business dealings, domestic life and care of the self.[29] The concept of character, understood as the individual who could discipline his impulses – he who could shape his conduct through the medium of aspirations – became a key definer of civilization.[30] In the words of Ann Stoler, the colonial gaze was 'a regard fixed on the colonized, but just as squarely on Europeans themselves.'[31] Thus, it was considered absolutely necessary for the British to remain in control of themselves to serve as models for the Ceylonese.[32]

As we shall see, the concept of moral hygiene – the medicalization of the passions – arose as part of a larger move towards sanitation for reasons of health and moral superiority both in Victorian England and the colonies.[33] As Stoler argues, 'European was not a fixed attribute, but one shaped by environment and class. It was contingent rather than being secured by birth.'[34] In other words, 'civilization' was something that was performative and embodied, in gestures, words, actions, and dress. The notion of European civilization was explicitly bourgeois. The poor and under-educated were seen as lacking the necessary degree of self-control, and hence were considered to occupy a position somewhere between recently-arrived, educated Europeans and mixed-race locals.[35] The ideal of the powerful European male was fostered in other ways as well. For example, colonial administrators were retired at 55 years old in order that 'no Oriental was ever allowed to see a Westerner as he aged and degenerated, just as no Westerner needed ever to see himself, mirrored in the eyes of the subject race, as anything but a vigorous, rational, ever-Alert young, Raj.'[36]

Housing for planters ranged from the primitive hut of the pioneer to large bungalows. Millie[37] describes his first home in the hills in the 1850s as a mud and wattle hut, approximately 20 feet by 12 feet, partitioned in half by a straw mat.

---

29 N. Rose, *Inventing Our Selves: Psychology, Power and Personhood* (Cambridge: Cambridge University Press, 1998), 33.

30 On this, see Dean, *Governmentality*, 11 and Wiener, *Reconstructing the Criminal*, 47.

31 Stoler, *Carnal Knowledge*, 1.

32 Gay argues that it was the idea of regulating the will and channelling energy that separated the Victorian bourgeoisie from other classes in Britain. P. Gay, *Pleasure Wars: The Bourgeois Experience. Victoria to Freud.* Volume 5. (London: Fontana, 1998), 19.

33 Wiener, *Reconstructing the Criminal*, 26.

34 Stoler, *Race*, 104.

35 For a discussion of the complexities of whiteness in colonial Barbados, see Lambert, *White Creole Culture* and in Brazil, see N.L. Stepan, *Picturing Tropical Nature* (London: Reaktion, 2001). A. Blunt, *Domicile and Diaspora: Anglo-Indian Women and the Spatial Politics of Home* (Oxford: Blackwell, 2005), 33, shows how in the early twentieth century a government commission in India chose not to distinguish between persons born in India of European parents and persons of mixed blood.

36 E. Said, *Orientalism* (New York: Vintage, 1979), 42.

37 Millie, *Thirty*. The term 'bungalow' was used even for a mud hut, as the term distinguished a European dwelling, no matter what its actual form, from a native dwelling. For

Figure 3.1   The new estate

Source: Hamilton and Fasson, *Scenes in Ceylon*, 1881.

One half consisted of a combined servant's room and kitchen, the other of the planter's bedroom and living room. There was a hole cut in the mat for the servant to come back and forth. Each room had a window and door, each with a plank on a leather hinge to shut it. The simple furniture consisted of a bed, a table and chair. After the first year on a new estate, a planter might construct a more substantial wooden dwelling consisting of a central living room, a kitchen and bedrooms off the living room. Such bungalows tended to be ill-furnished and ill-kept. Planters thought of them as transitional, a first step on the way to constructing proper homes for themselves and eventually for British wives.[38] Proper homes were thought to be key sites in the transformation of colonial society, representations of home in a foreign land. As Philippa Levine points out, 'English space was the polar opposite of colonial spaces, which were inevitably too crowded, hopelessly messy, loud and unordered. The contrast between a chaotic east and a well-regulated harmonious Britain was a potent one.'[39]

A key aspect of moral hygiene was proper housing. As one planter[40] put it, '[b]ad housing accommodation, after a time begins to tell on a man's general character ... When one gets dirty and careless in personal accommodation and appearance, worse is not far off ...' Central to this struggle was European, masculine, steadfastness

---

the definitive work on the bungalow as a colonial house type see A.D. King, *The Bungalow: The Production of a Global Culture* (Oxford: Oxford University Press, 1995).

38 On similar bachelor homes in India, see King, *The Bungalow*, 43.

39 Levine, *Prostitution, Race and Politics*, 297.

40 Millie, *Thirty*, n.p.

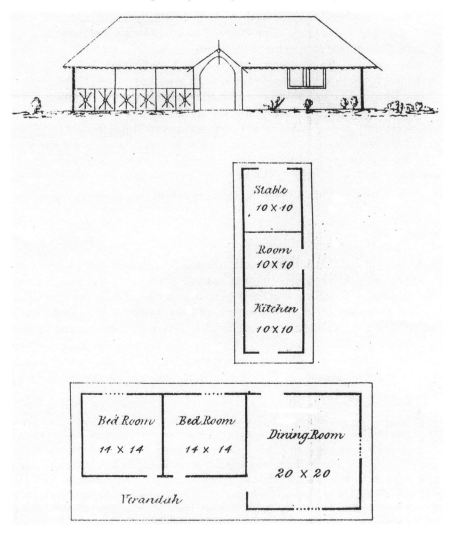

Figure 3.2   Plan of coffee planter's bungalow

Source: *PPA*, 1876–1877, p. 201.

and self-mastery.[41] The visible condition of slovenliness was thought to be both a sign of the onset of tropical disease and degeneration. Harrison notes that in Britain aristocratic standards of personal hygiene diffused to the middle classes by the mid-nineteenth century and that cleanliness of house and body had become a marker of respectability. As Bewell[42] puts it, '[p]eople are read through the landscapes they inhabit, while landscapes are seen as the physical expression—the topographical indexes—of the moral discipline and technological power of those who inhabit them.'

---

41 Rose, *Inventing Our Selves*, 31.
42 Bewell, *Romanticism*, 44.

Such identity markers were relational; as Mort[43] argues, 'bourgeois cleanliness was impossible without the image of proletarian filth, middle-class propriety could not be defined without the corresponding representations of working class animality ...' Increasingly during the nineteenth century, South Asians came to be seen as dirty, just as the poor in Britain were seen as the 'great unwashed.'

Europeans sought to regulate their diet, adjusting it to tropical conditions. They avoided exposing the body to the heat of the sun or taking too much exercise.[44] The humoural theory of disease lay behind such strictures for it was thought that in the heat of the tropics exercise would disrupt the humoural balance of the European body. Such concerns dictated the choice of clothing which included solar topees to cool the head, loose clothing, and flannel waistbands to absorb perspiration.[45] The practices of whiteness in the tropics were highly embodied.

Temptations

In the early days, plantations were widely dispersed and planters were unlikely to be accompanied by British wives. They had to battle against weakness brought on by loneliness and the temptations of the tropics. Brown,[46] in a locally published book aimed at the planter community, describes the planter as a 'modern day Robinson Crusoe.' The trope would have resonated with readers at that time, because the first plantations were thought of as widely interspersed islands of civilization in a vast sea of tropical nature and natives. Like servants in Britain, the natives were present physically, but not socially. Millie[47] in his reminiscences of planting in the 1840s writes:

> as far as the eye could see the horizon was bounded by this perpetual jungle..... A sense of utter loneliness came over me; it was worse than being at sea: the ship moves and we get out of the waste of water in time, but here is a fixture in the settled gloom of never-ending forest. Frequently, as evening approached, enveloped in thick mist, not a sound was to be heard but the sharp bark of the red elk, the scream of the night hawk, varied by the crashing of the elephants in the forest. The birds have no song during the day, insect sound is mute, and the silence can almost be felt. The moaning sound of the wind passing over the forests serves also to increase the feeling of gloom.

Millie[48] later describes the early plantations as prisons cut out of the forest. The images of the deserted island, the ship at sea and the prison are all images of a solitary life. And yet, there were many Tamil workers on the plantations and Sinhalese in the villages nearby. However, most early planters knew little Tamil or Sinhalese and

---

43 F. Mort, *Dangerous Sexualities: Medico-Moral Politics in England Since 1830* (London: Routledge and Kegan Paul, 1987), 41.

44 Harrison, *Climates*, 21–22.

45 Curtin, *Death By Migration*, 42–43.

46 S. Brown, *Life in the Jungle, or Letters from a Planter to his Cousin in London* (Colombo: Herald Press, 1845).

47 Millie, *Thirty*, n.p.

48 Ibid.

fraternising with male field hands was considered unacceptable. Household servants, who were nearly always male, occupied an intermediate status.[49] The planter might chat with them, but would not consider them friends. During the period of which I write, it was common to have a Sinhalese concubine in the bungalow, often with a room of her own.[50] But little or nothing was written of such mistresses who were thoroughly erased from the literature of the bungalow appearing nowhere in the planters accounts. They appear only in police accounts as a victim of planter violence or as the cause of violence between planter and a household servant.[51]

But even where there were other planters, loneliness could be a problem, especially for the newly arrived. For example, the 17-year-old James Taylor[52] expressed his home sickness in letters written shortly after he arrived from Scotland to assume his post as an assistant superintendent on a coffee estate in 1852. He often asked for more letters from home which, he said, 'would be a great comfort to me in this lonely wilderness'. And he missed the landscape of his native Scotland. He asked about the potatoes.

> I wish we had them here though they were diseased. I would give something for a peck of oatmeal too or a pint of milk—and have the cows been eating the corn? ... As yet I mind everything [at home] as distinctly as though I had left yesterday—every cut in the road and every large stone and all the blue hills and knolls.

There was a fear within the community and among families and officials at home that loneliness would lead planters to drink and the 'unhealthy association' with native men and women. While such concerns were normally couched in moral terms, both alcohol and consorting with native women were also thought to weaken the body, making it more predisposed to fevers.[53] Thus planters not only risked losing their Europeaness through miscegenation and degeneracy, they also risked an early death from disease.

Planter life was characterised by a sense of fraternity, masculine camaraderie, and homoeroticism. William Boyd[54] wrote that the planters in the 1830s and 40s were almost all bachelors who 'kept open house extended to all and sundry who

---

49 E. Carpenter, *From Adam's Peak to Elephanta* (London: Kessinger, 1910), 23, 75–80.

50 R. Hyam, *Empire and Sexuality: The British Experience* (Manchester: Manchester University Press, 1990).

51 Two cases appeared in the Annual Police Reports in the *Administrative Reports (ARC)* for 1880, Part 3, 26b. The first reported that 'a young coffee planter was shot at his dinner table with his own gun by his own servant, a Sinhalese youth about the age of his master. A few days previous, during the master's absence, the servant had made advances to a young Sinhalese girl whom the master "kept," for which he received a very severe thrashing. This rankled in his mind, and a few days later, after some additional annoyance during the dinner which he served to his master, the girl being present, he shot him dead from the passage. He was hanged.' The second case pertained to a planter who was 'accused of killing a Tamil woman—his mistress in this case also. He was acquitted, principally because the medical evidence showed sufficient disease to account for death.' Such was the typical path of racialised 'justice' in colonial Ceylon.

52 Taylor, *Papers*, 23 June, 1852

53 Harrison, *Climates*, 81.

54 Boyd, 'Ceylon', 266.

possessed a white face, who would conform to the habits of good society and possessed the rank of gentleman.' He went on to add that 'it was even extended to many who had, perhaps at one period possessed the rank of gentleman, but whose habits had excluded them from all the privileges of their order.' Anderson[55] refers to the social condition of colonialism as a 'middle class aristocracy' where issues of character rather than class background were paramount. Stoler,[56] however, argues that this varied from colony to colony and that previously fluid class distinctions had become rigid by the end of the nineteenth century. What Boyd suggests, however, is that the behaviour of some planters did not conform to proper metropolitan norms of respectability and manliness. They were outliers in the bio-political, statistical sense of the term, as well as the geographical. Their lapses were thought to be caused by several factors. The first was the tropical climate itself which weakened character. The second was isolation from positive civilizing influences such as the church. And the third was the relative absence of middle class wives.[57]

The colonial secretary the Third Earl Grey, expressed the commonly held view that subsistence in the tropics was so easy that people lost the will to work.[58] The impact of tropical nature, he argued, could only be overcome by inculcating in native people a belief in the 'dignity of labour.'[59] Furthermore, the energy-depleting tropics could only be resisted by those Europeans who had enough moral fibre to remain temperate. However, the issue of European labour in the tropics was complex; for although Europeans should avoid indolence, manual labour was considered beneath their dignity and more importantly physically dangerous for Europeans who were 'exotics in the tropics.'[60] Many of the planters' reminiscences were basically morality tales of how early planters were forced to do physical labour because they had no hired workers and how sometimes these efforts cost them their lives.[61] Their toil was portrayed as heroic, manly, self-sacrifice for the cause of progress. Such stories were contrasted with those of planters who sank into indolence and depravity. These accounts drew upon what Stoler[62] refers

---

55 B. Anderson, *Imagined Communities* (London: Verso, 1983), 137.

56 A.L. Stoler, 'Rethinking Colonial Categories: European Communities and the Boundaries of Rule', in *Colonialism and Culture*, edited by N.B. Dirks (Ann Arbor: University of Michigan, 1991), 119–152.

57 On this point see, I.M. Young, *Justice and the Politics of Difference* (Princeton: Princeton University Press, 1990), 136 and Blunt, *Domicile and Diaspora*, 26.

58 J. Tosh, 'Domesticity and manliness', in *Manful Assertions: Masculinities in Britain since 1800*, edited by M. Roper and J. Tosh (London: Routledge. 1991), 74, 92, notes that the Evangelicals moralized work and that Victorian manliness after mid-century came to be very much associated with work.

59 J. Cell, 'The Imperial Conscience', in *The Conscience of the Victorian State*, edited by P. Marsh. (Syracuse: Syracuse University Press, 1979), 193.

60 Arnold, *The Problem*, 160. It was thought, incorrectly, that medical topography demonstrated that biological race was an important factor in immunity and hence that certain races simply could not lead healthy lives in some locations (Curtin, *Death by Migration*, 47).

61 Millie, *Thirty* and Anonymous, *Days of Old: Or the Commencement of the Coffee Enterprise in Ceylon by Two of the Pioneers* (Colombo: A.M. and J. Ferguson, 1878).

62 Stoler, *Rethinking*, 125.

Figure 3.3   A hunting morning

Source: Hamilton and Fasson, *Scenes in Ceylon*, 1881.

to as the 'pioneering Protestant ethic' in which ultimate success results from hard work and perseverance. The most appropriate European work in the tropics was conscientious supervision.[63] In the words of Lewis, 'on many a mountainside, large tracts of scorched soil and charred timber bore evidence to labour directed by Anglo-Saxon energy.'[64] Such British energy was contrasted with the 'apathy and love of ease' of the local peasants whom the planters could rarely convince to work on the estates.[65]

Just as labour was structured by a set of ideals coded by race, class and degrees of masculinity, so was leisure. The main forms of recreation for the planters were eating, drinking and hunting with neighbouring planters.[66]

---

63 While this was the ideal, in fact many Europeans such as surveyors, engineers, government officials and planters worked hard in the tropics. However, such work was seen as distinct from manual labour. The British labouring classes were not encouraged to come to Ceylon as such work was thought to be more appropriately undertaken by indigenous peoples. Blunt, *Domicile and Diaspora*, 34, shows how in India there was concern about third-generation English settlers who were considered so degenerated as to be 'unemployable in any responsible position and unable to compete with Indians in manual labour.'

64 Lewis, *Coffee*, 9–10.

65 The image of the Sinhalese as lazy is particularly ironic because they did the hardest labour associated with the estates: the opening up of new forest land. The Sinhalese were willing to do piece work such as felling trees on plantations but were reluctant to commit themselves to long-term contracts, as these interfered with their own agricultural schedule. The Sinhalese were also unwilling to work for the low wages paid to the Tamils.

66 J. Weatherstone, *The Pioneers: The Early British Tea and Coffee Planters and Their Way of Life, 1825–1900* (London: Quiller Press, 1986), 176.

Some of the men, especially those from privileged backgrounds, played British country sports. They kept hounds and used native beaters to hunt elk.[67] Local commentators and older planters worried that too many of the young, privileged planters spent their time pursuing these types of entertainment rather than working hard on their estates. Again, it was the tropical environment that was seen to undermine the ambition to succeed.

Bachelor homes were also sites of ambivalent masculinity, in that they lacked a key element of bourgeois propriety, British women. On the other hand, as Tosh notes, around the mid-nineteenth century there was a certain shift in the meaning of masculinity such that young middle and upper-middle class men increasingly defined their masculinity against the domesticity of the metropolitan bourgeois home.[68] Seeking to escape this home space, which they increasingly defined as the realm of femininity, many British men looked forward to a life of bachelor adventures in the colonies. There also developed a cult of the freedom of masculinity, which was linked to the experience of the homo-social environment of public schools.[69] Rutherford argues that such environments produced a state of 'perpetual adolescence' among some members of the upper-middle and middle classes, whereby bachelor homes and clubs were thought of as an extension of boarding school.[70] Ambivalence surrounded this minority lifestyle, as it so clearly contradicted Victorian ideals of domesticity. Furthermore, the bachelor home was sometimes the site of illicit homo- and hetero-sexual relations and thus of shame, as evidenced by the erasure of such relations from their otherwise detailed accounts of planter life.

Drink and sexual relations with local people were activities about which planters as a whole were decidedly disturbed, at least in print. While it was admitted that planters drank, often to excess, drunkenness was bemoaned. Lewis[71] wrote of 'scenes of dissipation and extravagance' and 'absence of restraint' which 'seduced...[some planters] from their work.' Brown[72] condemned the weakness of some planters whose:

> [m]oral principles have not been strong enough to enable [them] to resist temptations to which a solitary life, distant from social amenities and religious restraints and privileges, has added force. Comfort is found in stimulants; the man takes to drink; that leads to habits and associations which deprive the victim of his own self respect and the respect of even his coolies it is his business to command.

These habits of the 'solitary life' were not usually described but they certainly entailed association with Tamils and Sinhalese including sexual liaisons with male or female workers or village women.

---

67 A Planter, *Ceylon in the Fifties and the Eighties*, 29.

68 Tosh, 'Domesticity', 68.

69 J.R. Mangan, *The Games Ethic and Imperialism* (Middlesex: Frank Cass, 1985); Tosh 'Domesticity'; Rutherford, *Forever England*.

70 Rutherford, *Forever England*, 23–24.

71 Lewis, *Coffee*, 14–15.

72 A. Brown, *The Coffee Planter's Manual* (Colombo: Ceylon Observer Press, 1880), 56.

Figure 3.4　A young Tamil labourer

Source: Royal Commonwealth Society Library.

The subject of women was even more delicate than that of drink. For while there were plenty of condemnations of the evils of drink, especially from those of a religious persuasion, it is rare to find any mention of sexual relations. By the mid-nineteenth century, the impact of the evangelical revival and the social purity movement made

British men even less open about liaisons with native women than earlier, and consequently references to sexual relations were guarded and often disapproving.[73] Having said this, until the 1860s, it was common for planters to have a native woman living in the planter's bungalow. Later, as sexual relations with native women became more disapproved of, women were kept in nearby villages.[74] Although household servants were considered 'fair game,' it was an unwritten rule that planters should not take Tamil women from the plantation work force as concubines, for this produced conflict with the male Tamil workers.[75] However, as one might expect, that rule was often broken, as suggested by a Tamil kangany by the name of Carpen who wrote of the frustration that male workers felt at the sexual harassment of their women by planters.[76] Consider Figure 3.4. Such photographs operated in code, for it was not acceptable for planters to keep pictures of young native women. And so while the photograph is ostensibly about work, it is an example of fabricated realism and, to borrow a turn of phrase from Alloula, it is simply 'brimming over with connotative signs.'[77] While not a studio shot, neither is it a typical estate shot. It is suspended somewhere in between the conventions of planting and those of photography. For the coffee bush appears to be cut and leaning against a wall in order to provide a background for the woman picking the berries. The photograph is atypical in other ways as well, for the majority of the labourers were men, but a young, attractive woman has been chosen as the subject. So while this photograph is ostensibly about the process of picking coffee, it appears to be more about a scopic desire. Furthermore, this photograph was taken during the years of collapse of coffee when, as we shall see in Chapter 7, the labour force on estates went hungry. This woman however, appears well-fed and dressed in new clothing rather than the rags that were the norm at this time. As such, the photograph not only speaks of the desirability of this young woman, but of the good life that estate labourers live.[78] In this sense, the photograph is pure ideology in the sense that Roland Barthes so skilfully developed in his *Mythologies*.[79] It resonates, through the image of the female body, of both carnal and economic desire.

At times the planter's sexuality was expressed in thinly repressed sado-masochistic terms. R.B. Tytler, seven-term President of the Planters' Association regularly flogged his male and female workers for infidelity.[80] But he also enjoyed

---

73 K. Ballhatchet, *Race, Sex and Class under the Raj* (London: Weidenfeld and Nicolson, 1980), 5.

74 Weatherstone, *The Pioneers*, 180.

75 Ceylon was particularly known for the availability of young boys as well as young girls as sexual partners. Hyam, *Empire and Sexuality*, 88, 95.

76 Moldrich, *Bitter*, 86.

77 M. Alloula, *The Colonial Harem* (Minneapolis: University of Minnesota Press, 1986), 21.

78 See Lambert, *White Creole*, 180 for some excellent examples of similar propaganda in early nineteenth century Barbadian depictions of plantation paternalism regarding well-fed, well-clothed labourers.

79 R. Barthes, *Mythologies* (London: Jonathan Cape, 1972).

80 Lambert, *White Creole Culture*, 62, notes that under slavery in the Caribbean, the whip was not only a technique of terror and discipline, but a symbol of white supremacy as well.

simulated rape scenes and would have the men capture a young female worker and drag her in front of him. Only if she put up a real fight would everyone get extra rations.[81] It appears that planters had a preference for girls in their early teens.[82] This was justified in terms of tropicality. Tropical peoples were thought to develop more rapidly than Europeans, especially sexually. Jenkins[83] writes '[i]n the east a girl is a woman at thirteen, a mother at fourteen, and a grandmother before thirty.' Boyd[84] echoes this when he writes, that native women were 'models of feminine beauty when in the first blush of their opening charms,' but 'by the time they reached the age when English women are at their best, their beauty has faded.'

It is fascinating to note that bourgeois British men applied these same ideas of the impact of the tropics on bodies and sexual drives to the lower classes in Britain as well. In the 1830s and 1840s, the Factory Commission described factories in Britain as artificially produced tropical environments, which had a physically and morally degenerative effect on workers. Gaskell wrote of:

[t]he stimulus of a heated atmosphere, the contact of opposite sexes, the example of lasciviousness upon the animal passions – all have conspired to produce a very early development of sexual appetencies [in the working class]. Indeed, in this respect, the female population engaged in manufactures, approximates very closely that found in tropical climates.[85]

Likewise, Engels[86] believed that, 'the factory has the same effect as a tropical climate, in both cases nature revenges herself on precocious physical development by premature old age and debility.' By the late nineteenth century, it was a commonplace to compare slums in Britain to the tropics, revealing, Edmond argues, anxieties about disease and degeneration.[87] Interestingly, the impact of the tropics on middle class

---

81 Moldrich, *Bitter*, 86. Tytler was one of the pioneer planters who came to Ceylon after working on a coffee plantation in Jamaica during the last years of slavery.

82 Apparently such attitudes were not restricted to the colonies for the age of female consent in Britain was only raised from twelve to thirteen in 1875. In the United States in the nineteenth century it was even lower, in most states it was ten and in Delaware it was seven! Hyam, *Empire and Sexuality*, 65–66.

83 A Planter, *Ceylon in the Fifties and the Eighties*, 12. This view of the early sexual development of women in the tropics was given authority by Montesquieu who wrote that girls were ready for marriage by eight years old and their beauty was finished by the age of twenty (Harrison, *Climates*, 101. Very occasionally a planter would admit to be in love with a local girl. Capper for example, writes of his love for a twelve-year-old Sinhalese girl, although he says that his love was never consummated (J. Capper, *Old Ceylon: Sketches of Ceylon in Olden Times* (London: W.B. Whittington, 1878), Ch. 1).

84 Boyd, 'Autobiography', 413.

85 P. Gaskell, 'The Manufacturing Population of England, Its Moral, Social and Physical Conditions, and the Changes Which Have Arisen from the Use of Steam Machinery' (London: Baldwin and Cradock, 1833), 68–69, cited in Bewell, *Romanticism*, 273.

86 F. Engels, *The Condition of the Working Class in England* (Oxford: Blackwell, 1958), 183. Cited in Bewell, *Romanticism*, 274.

87 R. Edmond, 'Returning Fears: Tropical Disease and the Metropolis', in *Tropical Visions in an Age of Empire*, edited by F. Driver and L. Martins (Chicago: University of

European women's sexuality was thought to be negative, taking the form of menstrual problems and reproductive dysfunctions.[88] In this class-specific interpretation of the impact of the tropics there is an odd, but on reflection understandable, refusal to consider the fact that middle class women might become eroticised by the tropics in the way that lower class white and native women were thought to be. Clearly bourgeois self-control would be particularly necessary for women in the tropics.

The ideal Victorian family home, seen by many as a key site in the spread of modern European civilization and thus essential to colonial development, was difficult to put into practice in the hills prior to the 1880s. In fact, as we have seen, the typical bachelor dwelling was nearly the antithesis of a proper home. In fact, it was common for planters to lament that after a number of years they were no longer able to participate in Victorian home life. In many cases colonial homes became primary sites of miscegenation and 'unproductive eroticism.'[89] Alison Blunt argues that because of notions of tropically-induced degeneration, the very idea of Victorian domesticity was problematic in South Asia.[90] The spectre of racial degeneration lay heavily on some of these men. For example, the formerly homesick James Taylor[91] wrote to his father after five years in Ceylon:

> now I am a confirmed bachelor and have not spoken to a European woman except once or twice for the last four years ... Faith I shall become as afraid of them as [Mr.] Pride [his supervisor] was. He never went near them and would even shut his eyes when passing a native woman.[92]

In another letter home he wrote, '[t]hough I would like well to get hold of a suitable wife, I don't know how. I can scarcely bring one from home as I have changed and been changed since I left and have so much to do with black people.'[93] We see in the highly restrained confessional quality to Taylor's writing that he felt he had been degraded by the tropics and his association with native women. He was no longer British in the full sense of the term and the domestic realm was unavailable to him. He continued in his letter of October 3, 1857,[94] 'I'll certainly never marry one [a native woman] and I don't know if a white woman would suit and a half-a-half is worse than either...striped.'

It is worth considering Taylor's letter in rather more detail. First, he expresses great ambivalence about white women, on the one hand expressing a desire to have a white wife and yet on the other expressing fear of white women and feelings of his own unworthiness, and suggesting at the end of his letter that any sort of wife, white, native, or mixed race, would be incompatible, being only a question

---

Chicago Press, 2005), 187–88.

88 Harrison, *Public Health*, 51.

89 The term unproductive eroticism is Stoler's (See Stoler, *Race,* 135).

90 Blunt, *Domesticity and Diaspora*, 36.

91 Taylor, *Papers*, 17 September 1857.

92 I would suggest that in shutting his eyes, he revealed his fear of his own desire for native women.

93 Taylor, *Papers*, 3 October 1857.

94 Ibid., 3 October 1857.

of which option was worse. And yet, while his relation to white women is one of fear, his relation to women of colour is one of aversion. He is no longer British enough for a British wife, but marrying a native woman would cast him altogether beyond the pale. As a mid-nineteenth century British man, Taylor was unprepared to take the step of marrying a native woman, as British men in South Asia during the eighteenth century routinely did. He considered the children of such a union 'striped,' the offspring of miscegenation. The issue of racial mixing, or hybridity as it was often termed was a matter of heated debate in the nineteenth century British colonies.[95] Degenerative views of race were propounded by ideologues such as Gobineau who believed that racial mixing would weaken the white race and strengthen the darker races. Even those of more liberal persuasion, thought that racial mixing could weaken both races.[96] However, many whites held an ambivalent attitude towards sexual relations with other races; white men were often attracted to native women, but their desires frightened them. Those of mixed race were both celebrated as objects of beauty and demonised as dangerous mutations of humanity. Reflecting further on James Taylor's letters home, one is led to ask to what extent his aversion to non-white women was a projection of his fear that Europeanness was slipping away from him under the impress of living in the tropics?

Kristeva's theorization of the abject may be helpful in understanding Taylor's state of mind. Abjection is expressed as feelings of disgust when it seems impossible to secure a firm boundary against an Other. Iris Marion Young[97]argues that racism, sexism and homophobia are all structured through abjection. As she says, 'the abject provokes fear and loathing because it exposes the border between the self and other as constituted and fragile, and threatens to dissolve the subject by dissolving the border.' She continues, 'the subject reacts to this abject with loathing as the means of restoring the border separating self and other.'[98] Ambivalent feelings of attraction and repulsion are experienced at a deep visceral level of bodily reaction, including aesthetic judgements. Beautiful Tamil girls in their early teens are seen as 'models of feminine beauty,' but by their late twenties they were said to be 'disgusting objects to look upon and repulsive beyond all imagination.'[99]

Men like Taylor felt even less at home in Britain, the place they still referred to as 'home'. For, 'going home,' meant to leave the Island, and return to Britain. But as one long time planter[100] who tried retiring to Britain put it, Britain has 'a harsh, cold uncongenial climate ... and an entirely new mode of life ...[which are] uncongenial.' He added that, when one returns to Britain, one is made to feel like a 'peculiar fellow, a sort of oddity, 'one who, you know, has lived all his life amongst black people, who is ignorant of [British] customs.' He concluded with a word of

---

95   On the history of hybridity see, R. Young, *Colonial Desire: Hybridity in Theory, Culture and Race* (London: Routledge, 1995).

96   Stepan, *Picturing Tropical Nature*, 109.

97   Young, *Justice*, 144.

98   Ibid., 145.

99   Boyd, *Autobiography*, 413.

100   Millie, *Thirty*, Ch. 27.

advice. If you go back, it is best to try and live in a hotel so that you can keep your freedom to be yourself. It is a familiar irony that it is in the *home* in Britain where those who return after years away may feel least at home. But the anxieties about returning 'home' to Britain after a long residence in the tropics were not related simply to one's cultural hybridity, but to physiological hybridity as well. The tropics were seen as having literally entered into the body and transformed it. Martin spoke of long-term residents of the tropics as 'tropical invalids' who could be healthy in neither tropical nor temperate environments. They were said to acquire a medical condition rendering them, 'climate struck' and 'morbidly sensitive to all the changes of season.'[101]

Bringing Women out: 'Angels for the Home'

In 1871 the barrister Richard Morgan travelled up to the hills where he observed the following:[102]

> It was a fine sight to see stalwart men assembled from miles around to join in divine worship. There was a harmonium to aid the singing which was very fine. There was one wont ... there were no ladies. The climate is fine and will suit Europeans. It would be a happy thing if each bungalow had a lady to adorn it, and planters would adopt the place as their home, and have each his family smiling around him.

Within a decade, Morgan's wish was beginning to be granted. Health conditions in the hills were improving and the death rates of the early decades greatly reduced. The 1869 opening of the Suez Canal reduced the length of the trip to Ceylon to a month and allowed white women, who were thought to have more delicate constitutions, to return 'home' more often. This led to increasing numbers of British women going out to South Asia. Whitehall encouraged women go as moral guardians of the home realm.[103] Metcalf argues that with the arrival of white women in South Asia, distinctions between the home and the world, at the heart of Victorian domestic ideology, were reinforced.[104] In fact, the colonial experience reinvigorated the ideology of public and private back in Britain.[105] In many ways South Asian women stood as the constitutive Other for British women. Progress was measured in part by the nobility and morality of the women. The higher the civilization, the more private, pure and domestic the women. British women would, it was felt, help create a more self-contained white society by 'rooting out' the native woman from the colonial

---

101   Martin, *The Influence of Tropical*, 461, cited in Bewell, *Romanticism*, 278.

102   W. Digby, *Life of Sir Richard Morgan Vol. 2.* (Madras: Addison and Company, 1879), 97.

103   Hyam, *Empire and Sexuality*, 119; Stoler, *Carnal Knowledge*, Ch. 3, reminds readers that the British, Dutch and French had all set in place regulations discouraging white women from emigrating to tropical colonies, on medical and financial grounds as well as due to a fear of miscegenation.

104   Hall, *White*, traces the history of a 'recodification' of British women as domestic beings to Evangelicalism in Britain during the early nineteenth century.

105   Metcalf, *Ideologies*, 94.

home. Their duty would be to create an atmosphere of rules, manners and restraint conducive to Christian manliness as defined by self-mastery, the ability, as Mosse[106] defines it, to restrain sexuality and other passions and impulses. During the last years of coffee, white women were thus actively recruited from Britain. The following is an excerpt from a letter to a London paper in 1884:

> There is just one word of advice I should like to give to fathers and brothers. To the latter, if you go to Ceylon arrange after you have a house of your own to get your sister out with you. England is overstocked with women, who are clamouring for votes and husbands too. Now England is sending out some of her best blood to its distant possessions. Why should the young men go and not the young women? I am convinced that the presence of his sister would have saved many a young fellow, in the pioneering days in the tropics, from drink and ruin, if she had been there to look after his bungalow and minister to his wants. Fellows used to come in from a hard days' work on the mountain slopes, fagged and weary, to their bungalow. There was food prepared for them by native servants but it was often not fit to eat. So some went to the beer or brandy for consolation. Things are better now and ladies are more numerous; but still, in colonising, whether to tropical or temperate climes, sister and brother may well go out together.[107]

While the presence of a white woman morally regenerated the colonial home, making it more respectable and bourgeois, her presence could never completely arrest environmentally-induced racial degeneration. An unbridgeable gulf between colonial homes and those of Britain was assumed to remain.[108]

The ambivalence of Britain-as-home can be seen in the reminiscences of a second-generation planter born in Kandy. He wrote:[109] '[a]ll this while my hopes lay on the prospects of going to England—"Home," – as I regarded England, though I had never seen the place!' Upon his arrival, he discovered that Britain would never be his home, thus his very self identity was called into question.

> Although I was born in an atmosphere of rigid British ideas; ideas that made me feel as thorough a Briton as if I had never been out of England, yet I lived in a climate that was tropical; surrounded with all the contracting influences, that unconsciously or subconsciously become Colonial; uneducated, untamed and with nothing to back me, pecuniary or otherwise, – these influences, in the aggregate, were bound to produce peculiar ideas of life, and angles of observation, completely different from those of our home-born countrymen.[110]

---

106 G. Mosse, *Nationalism and Sexuality* (New York: Fertig, 1985).

107 J. Ferguson, 'The Prospects of England's Chief Colony. An Interview with a Ceylon Journalist (Mr. J. Ferguson). *The Pall Mall Gazette* August 29, 1884', in *Ceylon and Her Planting Enterprise: in Tea, Cacao, Cardomom, Chinchona, Coconut and Areca Palms. A Field for the Investment of British Capital and Energy,* edited by A.M. Ferguson and J. Ferguson (Colombo: A.M. and J. Ferguson, 1885).

108 Blunt, *Diaspora and Domesticity,* 36.

109 F. Lewis, *Sixty Four Years in Ceylon* (Colombo: Colombo Apothecaries, 1926), 29.

110 Ibid., preface.

The feelings of inferiority to 'real' Englishmen welled up even more strongly later in his biography.

> All who knew me, knew that I was born in the country and utterly ignorant of English home life: and experience has often showed that the locally born was a far inferior creature; mostly despised, generally distrusted, and invariably classed as "country bottled," a particularly offensive appellation.[111]

As Hall points out, Englishness was, 'not a fixed identity but a series of contesting identities, a terrain of struggle as to what it means to be English.'[112] West Indian slave-owning planters earlier in the century had been seen as inferior types of British and in some cases barely British at all.[113] It appears that the fear that planters in mid-nineteenth century Ceylon had of being found wanting by metropolitan critics was in part a reaction to this older image of the tropical planter. Many of those who had left Britain years before or had never been there found that their ideas of proper British homes had become out-dated. In Britain, the Victorian ideal of domesticity and the gender relations it prescribed, had changed over the decades into a more fully developed and progressively more restrictive ideology with the separation of the home realm as feminine and the public as masculine having become ever more entrenched.[114]

Conclusion

In this chapter I have traced the way in which biopower in the form of internalised techniques of self-discipline and self care were confronted by what were seen as the challenges of a tropical environment and race. The passions, which were thought to be heightened by the tropics, had to be disciplined in the name of civilization and whiteness. As Stoler[115] argues, '[i]f we accept that "whiteness" was part of the moral rearmament of bourgeois society, then we need to investigate the nature of the contingent relationship between European racial and class anxieties in the colonies and bourgeois cultivations of the self.' The colonial home was idealised as a site of whiteness where the British domestic ideology of privacy, security and respectability could be nourished. However, Hepworth has argued that the middle class Victorian home in Britain should be conceptualised less as a haven and more as a battleground where bourgeois values struggled against the pressures of a dangerous world, a world that not incidentally included the ideology of colonialism already in crisis.[116] But if the metropolitan home was a battleground unable to defend its integrity against its constitutive outside – an imperial world in crisis, then the bachelor home in the

---

111 Ibid., 32.
112 Hall, *White*, 26.
113 Lambert, *White Creole*; Hall, *White*, 208.
114 Tosh, *'Domesticity'*.
115 Stoler, *Race*, 100.
116 M. Hepworth, 'Privacy, Security and Respectability: the Ideal Victorian Home', in *Ideal Homes?* edited by T. Chapman and J. Hockey (London: Routledge, 1999), 17–29; Gikandi, *Maps*.

tropics was ever so much more embattled. Even with the coming of British women, the home in Ceylon never provided the planter with a strong sense of physical security or the certainty of moral superiority necessary to fully legitimate his manly, Christian, imperial identity.

Chapter 4

The Quest to Discipline Estate Labour

Where there is power, there is resistance, and yet, or rather consequently, this resistance is never in a position of exteriority in relation to power.

Michel Foucault[1]

... coolies ... are mostly led by their canganies, and do as they like. We have considerable influence but no power over them; any power the law gives us it seldom enforces for us and coolies do not care for it.[2]

James Taylor, planter.

The master of a coffee estate is a small king in a small kingdom until he gets the sack... which proves to the coolies there is a hidden power still more supreme.[3]

P.D. Millie, planter.

Introduction

The coffee estates were unofficially recognised by the state as semi-autonomous spaces. A planter ruled his estate, in the words of P.D. Millie, as 'a small kingdom.' Planters represented themselves as practicing paternalism or what might be termed authoritarian governmentality in that they exerted control over all aspects of their labourers' lives, viewing them as useful resources to be managed in terms of their bodily capacities. However, much of the time, especially when the planters were hard-pressed financially, governmentality would be too benevolent a term to describe the regimes of practice on the plantations. As we will see here and in our discussion of health in Chapter 5, the estates were environments where it was very difficult for labourers to survive and planters did relatively little to aid them. One could say that the planters appeared to have as their goal, merely 'to make survive', to use Agamben's term.[4] And whether even that minimal level of care was achieved was often called into question by colonial administrators when they decided to intervene on the behalf of the migrant workers. One could thus describe the coffee estates as defacto sovereign spaces which sometimes came into conflict with the larger state over practices of governmentality.

In this chapter I examine some of the ways in which planters attempted to discipline their migrant labourers and to 'make them survive' or allow them to die.

---

1 Foucault, *The History of Sexuality: Volume I*, 95.

2 *PPA*, 1871–72, 263.

3 Millie, *Thirty*, n.p.

4 G. Agamben, *Remnants of Auschwitz: The Witness and the Archive* (New York: Zone Books, 1999), 155.

I also explore how they managed to enrol the state in their own economic projects. The estates were not 'spaces of exception' in the very strong sense that Agamben defines them. The state technically had authority over the estates. In effect, however, the estates' economic viability was so closely tied to the viability of the colony, that the government granted them a fair a degree of autonomy. The planters saw them as exceptional and tried to argue that they could most effectively and economically run their own affairs. The planters tried to discourage interference by the state in the lives of the labourers, when they thought that state legislation regarding migrant labourers would be costly for them, and encouraged interference when it would save them money. While planters did not have the right to take life, their decisions routinely created conditions that caused preventable deaths. Many of their actions would, in a court of law today, constitute manslaughter. However, as defacto-sovereign spaces, estates were spaces where the rights of migrant labour were often informally suspended. Although planters rarely directly took life, they knowingly allowed workers to die from neglect. This chapter explores the way the migrant labour population in Ceylon was governed as an economic resource for the estates and for the colony, which saw its interests as not identical, but closely tied to those of the planters.

How are we to know about the lives of poor workers living a century and a half ago? As Amin[5] points out, we often know them only through the violence inflicted upon them, for 'the speech of humble folk is not normally recorded for posterity; it is wrenched from them in courtrooms and inquisitorial trials.' Even more often, there were no trials and no confessions or testimonies wrung from them. In Ceylon, their bodies are simply used; drained of their powers and silently discarded by masters who were not interested enough to hear anything they might have to say. Where voices are mute, we attempt to learn from bodily practices and of these there remains little evidence. Furthermore, such evidence as can be found is often recorded by those with narrow and unsympathetic points of view. In this chapter, I tack back and forth in the fraught spaces between the practices of labour and the representation of those who all too often treated the labourers as disposable bodies.[6]

The plantations were constructed as what Henri Lefebvre has termed abstract space, purified space in which everyday life is thoroughly commodified and bureaucratised.[7] As an ideal, abstract space requires the construction of 'abstract bodies' which conform to it.[8] Abstract bodies, are bodies that are made docile,

---

5   S. Amin, *Event, Metaphor, Memory: Chauri Chaura 1922–1992* (Delhi: Oxford University Press, 1995), 1. Quoted in N. Wickramasinghe, *Dressing the Colonised Body: Politics, Clothing and Identity in Colonial Sri Lanka* (New Delhi: Orient Longman, 2003), 4.

6   For a useful discussion of representing 'the native side' see, D. Clayton, 'Imperial Geographies', 458.

7   H. Lefebvre, *The Production of Space* (Oxford: Blackwell, 1991). Although Lefebvre applies the notion of abstract space to capitalism in Europe, I would argue that it was equally a goal in industrial plantations in tropical colonies. There it was crosscut by race and south Indian social relations.

8   M. Landzelius, *personal communication*. See also M. Poovey, *Making a Social Body: British Cultural Formation, 1830–1864* (Chicago: University of Chicago Press, 1995).

useful, disciplined, rationalised, normalised, and controlled sexually;[9] such bodies are seen as economic resources to be rationally managed and utilised to capacity. In this respect the labourers were in effect indentured: a breach of contract was treated as a criminal offence and planters were allowed wide-ranging penal powers. What Behal and Mohapatra say of Assamese tea plantations applies also to Ceylon: 'It was the unfreedom inherent in indenture, namely the inability of the worker to withdraw his/her labour-power, to bargain over the terms of the contract or for higher wages, that led it to being viewed as a "new system of slavery."'[10] Such bodies operate under the gaze of authoritarian governmentality. The truly successful production of docile bodies in the colonial estates would ideally have required the creation of de-cultured workers.[11] Although the workers in question were poor Tamil villagers from southern India, who had been displaced from their traditional social networks: their social, judicial, religious, and medical support systems, they were never effectively de-cultured, just as the spaces of the plantations were in actuality never truly *tabula rasa*. The attempts to produce rationalised bodies for the estates were continually undermined as elements of the workers' networks and ways of life remained intact, serving as important bases for resistance and tactics of survival. Just as the harsh material realities of tropical nature strongly resisted rationalisation,[12] so the only partially-rent Tamil networks lent support to the workers' resistance to many of the attempts to re-culture them. Stoler argues that plantations failed to transform peasants into full-fledged proletarians.[13] Rather the plantation system allowed and even encouraged some elements of peasant life, such as the cultivation of private garden plots and traditional forms of authority. As such, the coffee estates in Ceylon can be understood as sites of heterogeneous networks, radically reconstituted into a new socio-material configuration, with some degree of economic stability and success achieved at a very high social cost, borne principally by the Tamil workers, but also by the planters themselves.

---

9 M. Foucault, 'Docile Bodies', in *Discipline and Punish: The Birth of the Prison* (New York: Vintage, 1979) 135–69.

10 R.P. Behal, and P. Mohapatra, 'Tea and Money Versus Human Life: The Rise and Fall of the Indenture System in the Assam Tea Plantations, 1840–1908', *The Journal of Peasant Studies* 19 (1992), 142.

11 Planters, as we shall see, viewed 'natives' as the product of a degrading climate and moral condition, which, it was believed, European capitalism could begin to transform. For a discussion of the discourse of the degraded 'native', see Arnold, *Colonising*; *The Problem*; Livingstone, 'The Moral', 'Tropical Climate'. J.S. Duncan, 'The Struggle to be Temperate: Climate and "Moral Masculinity" in Mid-Nineteenth Century Ceylon', *Singapore Journal of Tropical Geography* 21 (2000): 34–47; For a working out of some of these ideas in Britain, see Mort, *Dangerous Sexualities*, 25.

12 Duncan, 'The Struggle'.

13 A.L. Stoler, 'Plantation Politics and Protest on Sumatra's East Coast', *Journal of Peasant Studies* 13 (1986), 124; E. Wolf, 'Specific Aspects of Plantation Systems in the New World', in *Plantation Systems of the New World*, edited by A. Palerm and V. Rubin. (Washington: Pan American Union, 1959), 43, says that such plantation workers live 'double lives', while M. Tausig, *The Devil and Commodity Fetishism in South America* (Chapel Hill: 1980), 92, 103, 113, says they are 'limnal beings'.

I first examine some of the strategies and technologies adopted by European planters as they attempted to produce and regulate abstract space and bodies on coffee estates. I will argue that planters never achieved disciplinarity, as internalised control on the part of workers.[14] In part, this was because estate owners and managers saw workers as short-term, expendable, racialized, commodities so thoroughly embodied by their raw 'animal' nature and lack of exposure to European civilising forces that self-discipline was only minimally achievable.[15] Hence, rather than educating workers through any formal schooling or other programs of cultural training, the most that could be hoped for was an appreciation of 'the dignity of labour.' Instead, an authoritarian regime of discipline was practiced in which obedience was enforced at all costs. As such, most worker discipline had to be imposed through direct surveillance by Europeans and through the threat of physical force, based on the common environmental determinist view of race which assumed that only through coercion would tropical people work hard.[16]

I look at the refusal of workers to willingly submit to plantation discipline. I explore, as much as the extant records allow, what Scott has termed 'the weapons of the weak, that is tactics of resistance to the labour demands within the modern, experimental space of the plantation.[17] Unable to consistently instil discipline through anything other than constant surveillance and punishment, the planters sought more effective methods of discipline, which they believed required sovereign power within the bounds of the estate while at times enrolling the colonial state in the disciplining process but only when necessary. This brought the planters into conflict with both workers and their Tamil supervisors (*kanganies*), the networks of power and authority being highly complex and constantly evolving as modern industrial methods of plantation agriculture were developed. Conflict also arose at times with other bureaucratic forces such as the Ceylon, Indian and British governments and philanthropic groups who campaigned against the planters' continual resort to force. There were significant tensions, as we will see here and in subsequent chapters, between the sovereign claims of the planters and those of the state, and between what planters saw as the realm of economic freedom and the bio-political agendas of the state.

There have been a number of previous studies of domination and resistance on plantations. One of the earliest and best known is Genovese's discussion of

---

14 Foucault, 'Docile', L. Stewart, 'Louisiana Subjects: Power, Space and the Slave Body', *Ecumene* 2 (1995): 227–45 explores New World slave plantations in much the same terms.

15 The struggle that white male planters waged against their own embodiment in the tropics is the subject of Chapter 3.

16 Arnold, *The Problem*, 160.

17 Scott, *Weapons of the Weak*. Like Steward, 'Louisiana subjects', I am keen on de Certeau's use of the term tactics, to refer to the manner in which the oppressed 'poach' in the space of their oppressors, but like her, I am uneasy with his claim that the oppressed can only resist through controlling time and not place. See M. de Certeau, *The Practice of Everyday Life* (Berkeley: University of California Press, 1984). Having said this, I will use de Certeau's notions of tactics to refer to the workers' resistance and strategies to refer to the planters attempts at domination.

everyday resistance on American slave plantations.[18] Researchers have found that trying to discern covert resistance from colonial records is extremely difficult. Ramasamy, after outlining some forms of resistance on colonial coffee plantations in late nineteenth century Malaya, concludes that 'lack of evidence prevents us from providing a detailed picture of individual forms of resistance among Indian labourers on plantations.'[19] Perhaps Warner has undertaken the most sustained set of studies of migrant Indian plantation labour.[20] Her study of the plantation industry in Mauritius in the nineteenth century provides a detailed understanding of the workings of many aspects of plantations and the recruitment of labour. However, she has relatively little to say about labourer resistance,[21] presumably for the same reasons as Ramasamy. One of the most successful at uncovering worker resistance is Behal, who in a series of studies of Assam tea plantations in the nineteenth and early twentieth century, documents the struggles over labour practices.[22] Speaking in more general terms, Adas[23] argues that non-confrontational resistance on plantations can be divided into what he terms everyday resistance, exit strategies such as desertion, and retribution such as damaging crops and equipment. He claims that everyday resistance is the most common form and the most difficult to study, because it involves individual rather than collective action and because it entails 'calculated errors' and incompetence rather than sustained protest. The principal goal of this chapter is to provide a detailed account of the forms of everyday resistance employed by estate workers and the strategies planters adopted to combat them. I use the term resistance with some trepidation here, for as Scheper-Hughes aptly notes, the goal of poor peasants is usually 'not *resistance*, but simply *existence*.'[24] Suffice it to say that I conceptualise resistance, not as a formal political act, but as a set of tactics adopted by the poor in order to make existence bearable and or even just possible in an environment unconducive to survival.

---

18 M. Genovese, *Roll Jordan Roll: The World the Slaves Made* (New York: Random House, 1972), 285–324, 599–657.

19 P. Ramasamy, 'Labour Control and Labour Resistance in the Plantations of Colonial Malaya', in *Plantations, Proletarians and Peasants in Colonial Asia*, edited by E.V. Daniel, H. Bernstein, and T. Brass (London: Frank Cass, 1992), 100.

20 M. Carter, 'Strategies of labour mobilisation in colonial India: the recruitment of indentured workers for Mauritius', in *Plantations, Proletarians and Peasants in Colonial Asia*, edited by E.V. Daniel, H. Bernstein, and T. Brass (London: Routledge, 1992), 229–45; M. Carter, *Servants, Sirdars and Settlers: Indians in Mauritius 1843–1874* (Delhi: Oxford University Press, 1995).

21 Carter, *Servants*, 225–226.

22 R.P. Behal, 'Forms of Labour Protest in the Assam Tea Plantations, 1900–1930', *Economic and Political Weekly* 20 (1985): 19–26; Behal and Mohapatra, 'Tea and Money.'

23 M. Adas, 'From Foot Dragging to Flight: the Evasive History of Peasant Avoidance Protest in South and Southeast Asia', *Journal of Peasant Studies* 13 (1986): 64–86.

24 N. Scheper-Hughes, *Death Without Weeping: The Violence of Everyday Life in Brazil* (Berkeley: University of California Press, 1992), 533.

Barnett[25] and McEwan[26] have rightly suggested that academics overstress 'voice' as a key element of resistance on the part of the subaltern. This is very much in line with the findings of Scott, whose theory of the weapons of the weak is premised on the fact that the oppressed usually cannot afford to let their voice be heard. All three conclude that a focus upon practices in understanding resistance is more productive. This chapter explores resistance as a set of embodied practices. Butler has argued that the body is a 'site of incorporated history.'[27] The Tamil workers' bodies on coffee estates can be seen as sites of memory and including importantly Hindu beliefs and practices brought with them from their villages in India. In fact, Hinduism is particularly focused on the body, having many bodily prescriptions and prohibitions associated with the practice of caste. The estates, then, can be seen as sites in which the incorporated histories of Indian villagers were often violently confronted by attempts to institute a modern European bio-political regime.[28] Although planters did not expect to produce fully modern individuals with interiorised self-control, they did hope to accomplish more limited goals through disassociating the labourers at least partially from their Hindu cultural incorporation. Through modern work discipline, it was hoped that culture change would gradually take place.

The Recruitment Process

The early planters tried to recruit Sinhalese peasants from surrounding villages. While planters convinced them to clear land on short term contracts and to occasionally work on a longer term basis as estate labourers, it soon became clear, that the Sinhalese were not willing to supply the necessary labour for estates.[29] There were

---

25 C. Barnett, 'Sing Along with the Common People: Politics, Postcolonialism and Other Figures', *Environment and Planning D. Society and Space* 15 (1997): 137–54.

26 C. McEwan, 'Cutting Power Lines Within the Palace: Countering Paternity and Eurocentrism in the Geographical Tradition', *Transactions, Institute of British Geographers*, 23 (1998): 371–84.

27 J. Butler, 'Performativity's Social Magic', in *The Social and Political Body*, edited by T.R. Schatzki and W. Natter (New York: Guilford, 1996), 30. For a similar point with regard to social memory see, P. Connerton, *How Societies Remember* (Cambridge: Cambridge University Press, 1989).

28 The concept of embodiment, the idea that all humans are thoroughly embodied, is valorised in contemporary social thought and is seen to dissolve entrenched dualisms between human and non-human, male and female, sex and gender, mind and body. In the nineteenth century, the racist, imperial view saw natives, women and the lower classes as particularly embodied. I will therefore attempt to use this concept carefully in order to separate these two rather different notions of embodiment.

29 Meyer argues that the role of Sinhalese labour has often been underestimated. In 'Enclave', 207, he claims that some level of employment of Sinhalese peasants 'was important to the economic processes of both the villages and the estates, at least during certain periods and in certain areas'. Nevertheless, even Meyer himself does not deny that only a relatively small percentage of plantation labourers were Sinhalese, especially after the initial clearing work was completed. In fact he calculates that in 1871 when coffee production was near its peak, the Sinhalese only formed slightly more than 3 per cent of the plantation workforce.

various reasons for this. The Sinhalese population had declined during the brutal British suppression of a rebellion in 1818–19. Many people had either been killed or had died from disease due to the disruption of water supplies which caused a severe outbreak of dysentery. Continuing high death rates in the second quarter of nineteenth century due to disease greatly reduced the available labor pool. Peasant agriculture was generally sufficient for this reduced Sinhalese demand. Working for the British was seen as an unattractive option because the rates of pay were low and the working conditions poor. The planters were forced to turn to more distant sources, first the coastal lowlands and then across the Palk Strait in Madras.[30] Like the Kandyans, the lowland Sinhalese had little interest in labouring on the estates, and planters complained that they often headed home for two or three months to fulfil 'social obligations' at times when they were needed for planting or picking the crop.[31] So the planters turned to Indian Tamils.

The Migration of Tamil Labourers

In the early years of the nineteenth century Sinhalese labour on public works was supplemented by small numbers of labours from South India.[32] Governor Barnes and Lt Colonel Bird were the first planters to import Tamils, bringing 150 labourers over to work their plantations at 15 shillings per man per month during the 1820s. However, the experiment failed due to the harsh treatment meted out on the estates. All who survived returned to India within the year.[33] Nevertheless, South Indian Tamils remained an attractive option for the planters, and systematic recruitment began in 1839 coinciding with the beginning of the boom in coffee production.[34]

The Tamils were willing to come for a number of reasons, some directly related to the British control over Madras, others due to relations with local south Indian elites, and still others due to weather. First, the local population of Madras grew as a result of a British-imposed peace in the area which produced an increase in the

---

See E. Meyer, 'Between Village and Plantation: Sinhalese Estate Labour in British Ceylon', in *Asie du Sud: Traditions et Changements* (Paris: Colloques Internationaux du CNRS, 1979), 460. On this point also see, R. Wenzlhuemer, 'The Sinhalese Contribution to Estate Labour in Ceylon, 1881–1891', *Journal of the Economic and Social History of the Orient* 48 (2005): 442–58.

30 *Ceylon Observer*, 10 July 1843. An early attempt was also made to bring in labourers from China but this was quickly abandoned on the grounds that it was too expensive and they were not sufficiently docile.

31 *Ceylon Observer*, 3 June 1843

32 *Sri Lanka National Archive* (*SLNA*) 5/2 North to Hobart, dispatches 3 October 1804 and 5 October 1804.

33 C.R. De Silva, *Ceylon Under the British Occupation 1795–1833*. Volume 2. (Colombo: The Colombo Apothecaries' Company, 1962), 484.

34 The average number of annual arrivals for 1839–42 was 5,300. This rose to 115,416 between 1873–79 and then dropped to 46,152 between 1880–86 when coffee was in its final decline. Bandarage, *Colonialism*, 202. From the middle of the nineteenth century Indian migration increasingly was restricted to Burma, Ceylon and British Malaya. Between 1834 and 1900, 92.2 per cent of emigrants went to this region. Kurian, *State*, 13.

Table 4.1   Immigration and emigration 1839–1890 (average #/year)

Year	Immigration	Emigration	Net
1839–42	5,300	6,900	-1,600
1843–50	47,028	19,693	27,335
1851–60	57,464	31,443	26,021
1861–70	68,415	53,185	15,230
1871–80	102,511	82,471	20,040
1881–90	57,856	52,752	5,105

Source: Adapted from Kurian, *State*, 62; H. Marby, *Tea in Ceylon* (Wiesbaden: F. Steiner Verlag, 1972), 27.

number of landless labourers. Second, poverty in the region was exacerbated by the decline of village weaving, as Indian cloth manufacture could not compete with the British cloth which had flooded the market.[35] Third, as Gough's study of Thanjuvur, an area of seasonal migration to the coffee plantations, demonstrates, the freeing of slaves in the south of India in the 1830s created a poor low-caste group in search of employment.[36] This coincided with the interests of local Indian landlords. Kurian[37] following Gough argues:

> [w]hen they were needed for cultivation in their villages, the landlords operated their authority of debt bondage to retain them in India. When they were not necessary for that purpose, it was also in the interests of landlords to allow them to work on the plantations in Ceylon. The workers, caught between the dictates of plantation capitalism and debt bondage, had little choice of place of work or seasonality of employment.

This group of landless labourers was pushed and pulled along networks joining India and Ceylon by the fluctuating demands of the coffee and rice crops. Fourth, the seasonality of cropping and fluctuations in weather patterns affected migration. As Kurian[38] points out, the demands of rice cultivation fell during the southwest monsoon (June–September) and the northeast monsoon (October–March) with harvesting in January and March. Labour was needed on coffee plantations between August and November with the greatest demand during the gathering of the crop in November. Sometimes, labourers could participate in the coffee harvest in November and return to India for the rice harvest in January. However, if the rains were delayed then problems of timing would arise. Also if the rice harvests were good in India, workers would feel less need to come to work in Ceylon and planters suffered shortages of

---

35 C. Kondapi, *Indians Overseas 1833–1949* (New Delhi: Oxford University Press, 1951), 2–4.

36 K. Gough, *Rural Society in Southeast India* (Cambridge: Cambridge University Press, 1981).

37 Kurian, *State*, 58.

38 Ibid., 59.

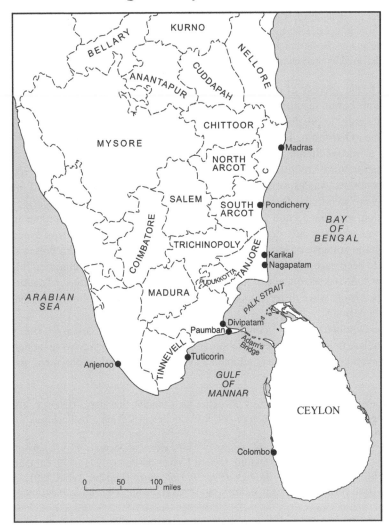

Map 4.1　South Indian sources of plantation labour

Source: Adapted from I.H. Vanden Driesen, *Indian Plantation Labour in Sri Lanka: Aspects of the History of Immigration in the 19th Century* (Nedlands: University of Western Australia, 1982).

labour. Conversely, in a year when the monsoons were insufficient or failed to arrive, labourers flocked to the plantations in Ceylon.

The labourers often arrived on the plantations in a state of collapse from famine and disease. The Government Agent of Jaffna Province, W.C. Twynam, described the migrant labourers of the mid-1840s as:

> miserable gangs of coolies... with one or two women to 50 or 100 men, strangers in a strange land, ill-fed, ill-clothed, eating any garbage they came across (more however from necessity than choice), travelling over jungle paths, sometimes with scarcely a drop of

water to be found anywhere near them for miles, and others knee-deep the greater part of the way in water, with the country all round a swamp; working on estates just reclaimed from jungle, or on jungles about to be converted into estates, badly housed, and little understood by their employers...[39]

As such, the success of coffee in Ceylon not only depended on local agricultural conditions but on the monsoons and their impact in south India. Such a dependency made it nearly impossible to predict the quality of a harvest or the amount of labour available to harvest it. This, in turn, made coffee a risky investment.

As a rule of thumb, labourers were willing to come to the estates in Ceylon when agricultural conditions were poor in southern India. The entanglement of the two economies was especially obvious when the south of India suffered famine.[40] There were two major famines in the Madras Presidency during the coffee period in Ceylon, in 1853–55 and 1876–78. While the principal cause of famines in south India was either a lack or overabundance of rain, the situation was exacerbated through the collapse of local industry, changes of land tenure, and a failure of local wages to keep pace with price increases.[41] There were surges in emigration following famines.[42] As a consequence of the social-environmental conditions in Madras, there was a persistent wage differential between Madras and Ceylon throughout the coffee growing period.[43]

The labourers were mainly drawn from the lower castes.[44] The planter Sabonadiere noted that low caste labourers were preferred as they were thought to be harder workers.[45] The planters adopted a system of seasonal employment, using the

---

39 *Colonial Office (CO) 54/475*, Letter to Colonial Secretary, Henry T. Irving, in 'Correspondence on the condition of Malabar Coolies in Ceylon', 16 Enclosure no. 8, 30 September 1869.

40 Snodgrass, *Ceylon*, 24; H. Tinker, *A New System of Slavery: The Export of Indian Labour Overseas 1830–1920* (Oxford: Oxford University Press, 1974), 119. On the Madras famine of 1876–78 see M. Davis, *Late Victorian Holocausts* (London: Verso, 2001).

41 B.M. Bhatia, *Famines in India: 1860–1965* (London: Asia House Publishing, 1963), 16–21.

42 D. Kumar, *Land and Caste in South India: Agricultural Labour in Madras Presidency during the Nineteenth Century* (Cambridge: Cambridge University Press, 1965).

43 Wages in Madras remained at three pence per day until the 1850s, while a male labourer in Ceylon could earn four pence per day in the 1840s and between six pence and nine pence per day in the 1850s. *Ceylon Observer*, 20 April 1843; *Ceylon Examiner*, 16 June 1852. For a general discussion of Indian labour on plantations see, M. Roberts, 'Indian Estate Labour in Ceylon During the Coffee Period 1830 to 1880', *The Indian Economic and Social History Review* 3 (1966), 3.

44 The Planters' Association *(PPA*, 1871–1872, 162–63) conducted a survey in the early 1870s and found that 72 per cent of their workers were lower caste.

45 W. Sabonadiere, *The Coffee Planter of Ceylon* (Guernsey: MacKenzie, Son and Le Patourel, 1866), 85; Kurian notes that numerically, the dominant castes on estates were sudra and adi-dravida. The former were low to middle and the latter were low. (see, R. Kurian, 'Labour, Race and Gender on Coffee Plantations in Ceylon (Sri Lanka), 1934–1880', in *The Global Coffee Economy in Africa, Asia and Latin America, 1500–1989*, edited by W.G. Clarence-Smith and S. Topik (Cambridge: Cambridge University Press, 2003), 184.

kangany system of recruitment. *Kanganies* were paid a head-fee to bring labourers from the south of India;[46] out of this fee they were to feed them on the march to the plantations. This cost was then deducted from the labourers' wages after they arrived. The *kanganies* remained on the plantations as gang bosses. Gangs varied from 25 to 100 labourers.[47] Heidemann divides the *kangany* system in Ceylon into two periods, pre- and post-1850.[48] During the late 1830s and 40s, many *kanganies* came from the labouring classes. They recruited their own relatives and others from their home villages. It may be reasonable to assume that during these decades *kanganies* felt a degree of solidarity with their labourers. After 1850, with the takeover of many plantations by institutional investors, *kanganies* began to fill the growing demand for labour by recruiting outside their own villages. As a result the relations between *kangany* and labourer became increasingly divorced from kinship ties.

A system called Coastal Advance was instituted in the 1850s. Under this system, *kanganies* were paid one pound a month as a retainer and one to two shillings per head for each worker that they brought over. It was up to the kanganies to recover the head money from the labourers. Workers at this time usually stayed for five to twelve months and then returned with whatever savings if any they were able to accumulate.[49] The small remittances that were returned to the villages in south India were often used to repay debt there.[50] Although not officially an indenture system, in fact all Tamil labourers arrived in Ceylon in debt to the *kanganies* and were not allowed to leave the estate until their *tundu* (note of indebtedness) was cleared. As Bandarage points out, in fact 'the labourer had moved from debt bondage in his South Indian village to debt bondage [to the *kangany*] on the Ceylon plantation.'[51] A kind of carceral continuum was established between India and Ceylon, with guards responsible for keeping labourers on their estates and, when that failed, government police recruited to round up runaways. The result of all this was a complex circulatory network established by planters and *kanganies*, which connected villages in Madras to the coffee estates in the Kandyan highlands.

The system of labourer recruitment had many advantages from the point of view of the planters. First, it delegated recruiting to a third party. Second, given the seasonal

---

46 A large majority of the labourers for Ceylon was drawn from the Madras Presidency, mainly from the districts of: Madura, Tinnevelly, Ramnad, Trichinopoly, Salem, South Arcot, Chingleput, Tanjore, Mysore, Malabar, and Travancore. The first wave came from Malabar and throughout the remainder of the coffee period the labourers were called Malabars, even though few came from there by the end of period. Vanden Driesen, *Indian Plantation Labour*, 7.

47 Kurian, 'Labour', 180.

48 F. Heidemann, *Kanganies in Sri Lanka and Malaysia: Tamil Recruiter-cum-Foreman as a Sociological Category* (Munich: Anacon, 1992), 12–13.

49 Vanden Driesen, *Indian Plantation Labour*, 52.

50 Bandarage, *Colonialism*, 210.

51 Ibid., 204. In Madras, peasants (Ryots) were hit hard by the over-assessment of their lands. Taxes of up to one third of the value of the crop had to be paid in cash annually. This drove peasants to moneylenders and a crushing cycle of debt. While the peasants lived hand-to-mouth, the landless labourers who constituted the bulk of the labour for Ceylon plantations were often destitute and in a state of debt bondage. H. Chattopadhyaya, *Indians in Sri Lanka* (Calcutta: OPS Publishers, 1979), 21–22.

demands of coffee planting, workers were hired only as needed.[52] Third, workers paid their own way to the estates and fourth, the labourers were far away from home. Being largely disconnected from their support systems and kept within the relatively well-circumscribed, carceral space of the estate, they were less difficult to control than low-country Sinhalese who could more easily return to their own villages.[53] Authoritarian biopower operates more effectively within such a prostrate civil society, although as we shall see, the workers were able to exert sufficient covert power to disrupt many of the strategies of the planters.[54] The recruitment system reinforced itself once the majority of the labour force was Tamil, for it became even more difficult to recruit Kandyan and low-country Sinhalese peasants. The local press at the time noted with regret that, the Sinhalese refused to work with low caste Tamils and were especially reluctant to work under the alien regime of Tamil *kangany* overseers.[55]

Throughout the coffee period, the majority of labourers who came to work on the estates were male. Planters sought to recruit women and their numbers increased steadily over the decades from 2.6 per cent in 1843 to 20.5 per cent in 1878. They did so because women were paid at a lower rate than men and thus their labour was considered more economical for many tasks. Women were also increasingly recruited because they were believed to be more reliable and docile than men. Further, it was also thought that a man with a wife and children would be more domesticated and likely to remain on an estate throughout the full season. Finally, and not insignificantly, they were brought to 'satisfy the sexual appetites' of the male workers, and keep them from leaving the confines of estates where they might come under the influence of bazaar keepers who sold them liquor often on credit. The bio-political ambitions on the part of the estate managers with the cooperation of the state here is evident.[56]

Transportation

During the coffee period there were two principal routes by which the labourers arrived at the estates. In the late 1830s and 1840s, labourers made the five-hour journey across the Palk Strait in small boats from Ramnad to Talaimanaar. From there the labourers faced a 150 mile walk to the estates. The trip from southern India to the estates normally took between seven and nine days.[57] In the first decade the coastal route from Arippu south to Puttalam and then east to Kurunegalla and on to

---

52 For a discussion of similar employment practices in twentieth century Californian agriculture see D. Mitchell, *The Lie of the Land* (Minneapolis: University of Minnesota Press, 1996).

53 E. Meyer, 'The Plantation System and Village Structure in Rural South Asia', in *Rural South Asia: Linkages, Change and Development*, edited by P. Robb (London: Curzon, 1983), 40.

54 For a more general discussion of the link between authoritarian modernist experiments and a prostrate civil society, see Scott, *Seeing Like a State*.

55 *Ceylon Observer*, 3 June 1843; *Ceylon Observer*, 5 January 1852.

56 The French unsuccessfully attempted to introduce women to their penal colony on Devil's Island for similar biopolitical reasons. See Redfield, 'Foucault in the Tropics', 64.

57 D. Wesumperuma, *Indian Immigrant Plantation Workers in Sri Lanka: A Historical Perspective 1880–1910* (Kelaniya: Vidyalankara Press, 1886), 44.

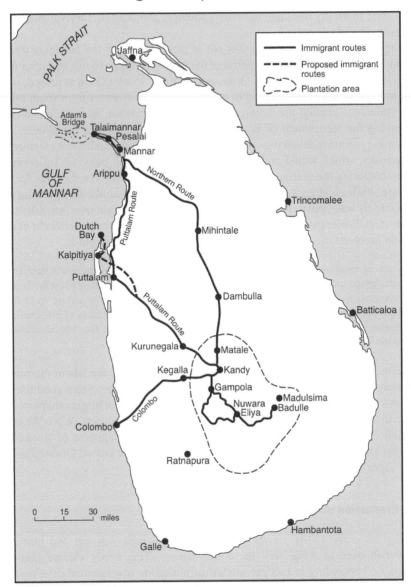

Map 4.2   The routes to the estate region

Source: Adapted from I.H. Vanden Driesen, *Indian Plantation Labour in Sri Lanka: Aspects of the History of Immigration in the 19th Century* (Nedlands: University of Western Australia, 1982).

the Kandyan planting districts was most often used. After 1850, the northern road to Mihintale, Dambulla and the planting districts was used more frequently.[58] Both of these became 'roads of death' for large numbers of labourers. I will examine the

58 Vanden Driesen, *Indian Plantation Labour*, 7.

fluctuating mortality rates and the state's attempts at managing survival along these routes in detail in Chapter 5.

During the late 1850s a plan was put in place to reroute the labourers through Colombo and thereby avoid the perils of the overland route. To this end after much lobbying on the part of planters, Ordinance Number 15 of 1858 was passed, which sought to provide government aid in the recruitment of workers from south India.[59] The Ordinance created the Immigrant Labour Commission to look into ways of improving the recruitment of labour and subsidize the costs to the planters. The principal recommendation was to have the government purchase a steamer, the Manchester, which would make two to five trips a month between Tuticorin and Colombo during the season. Men and women were charged three shillings for their passage, while children were charged between one and two shillings depending upon their size. It was hoped that 45,000 workers could be brought over annually using this system. While paying lip-service to laissez-faire values, the government became heavily involved in subsidising the estates. As Vanden Driesen put it,

> Short of meeting the entire cost of the scheme, the government could hardly have participated more closely than it now proposed to do. It was represented on the Board of Immigration Commissioners; it was financially responsible for the setting up of labour depots in Colombo and upon the Indian coast; it was to bear the costs of purchasing two steamers and it was to aid in overcoming the actual problems of the scheme with a loan of sterling 20,000.[60]

Nevertheless, this effort by the government to help manage the labour recruitment and transportation process foundered, for in the following two years good harvests and new public works projects in Madras attracted labour that might otherwise have come to Ceylon, and the ship ran at a huge loss to the government. When the newly appointed Governor MacCarthy arrived he was a keener exponent of laissez-faire than his predecessor, Governor Ward, and consequently he passed Ordinance 15 of 1861 repealing the immigration scheme.[61]

The Production of Abstract Space

During the early 1840s, newly expropriated forestland was cut and burned at a feverish pace to make way for coffee bushes. This newly cleared land was rationalised according to modern models of plantation management developing in India, Java, Jamaica and elsewhere.[62] In fact, estates in Ceylon from the 1830s on

---

59 Ibid., 70.

60 Ibid., 70.

61 Ibid., 71, 73.

62 Brown, *The Coffee*. As noted previously, Governor Barnes moved the Botanical Gardens from Colombo to Peradeniya adjacent to his newly opened plantation, precisely so that he could avail himself of the latest scientific advice on coffee production. The Ceylon Agricultural Society was formed in 1842, in order to improve agriculture, organise the introduction of Tamil labourers, form a library and correspond with other colonies about agricultural matters. The Society's first proceedings published a list of plantations

were organised on what was termed the 'West India system of cultivation.'[63] This putting into practice in Ceylon of spatial models of plantation management and worker control developed elsewhere and learned from texts rather than practical local experience was problematic for a number of reasons. First, Ceylon's environment (soil, weather, and pests) differed from other coffee-growing regions, and second, and more importantly, most of the owners, supervisors, and labourers on the new estates had never grown coffee before. With the price of coffee high in the early 1840s and with a few estates making quick profits, speculators rushed in buying land at inflated prices and hiring inexperienced staff. When the price of coffee collapsed in 1847, the ensuing debts were enormous. This collective failure was put down not only to a depression in Europe, but also to greed on the part of speculators. The Governor Lord Torrington summarised his view of the reasons for the collapse to the Colonial Office thus:[64]

> Soldiers whose discharges were purchased from the ranks were sent up to the interior to manage plantations on salaries of £300 to £400 per annum; houses for their use were purchased by the agents at excessive prices, and their style of living, wines, and their expenditures of every description were on a scale of the most absurd proportions, while the proprietors were mortgaging every security and raising money at nine or ten per cent to support this expenditure, buoyed up by the confident expectation that the first golden harvest would reimburse every outlay and leave them in possession of a splendid and permanent income.

While the aristocrat Torrington, put the problem down to the wastefulness of the lower classes who put on airs and took advantage of their superior's greed, a planter of more modest background writing in the *Ceylon Examiner* explained it as a loss of rationality which he compared to an addiction; 'an infatuation appears to have possessed them [the European planters], and they slumbered on like the narcotised opium smoker who having filled his lungs with the fumes of the pernicious drug, throws himself back on a couch.'[65] The implication of such critiques is that the early planters failed to behave as proper British imperialists.[66] Rationalised, textualized, abstract space could not be produced under such conditions.[67] What was needed to

---

and the number and wages of workers employed. *Proceedings of the Ceylon Agricultural Society* (Kandy: 1842). In the following year, the Society communicated with the Calcutta Agricultural Society, and with agricultural societies in Egypt and the Cape. *Proceedings of the Ceylon Agricultural Society* (Kandy: 1843). Some of the early planters such as R.B. Tytler had previous experience planting coffee in Jamaica (J. Ferguson, *Pioneers of the Planting Enterprise in Ceylon Vol. 1.* (Colombo: A.M. and J. Ferguson. 1894), 14.

63 J. Ferguson, *Ceylon in 1893* (London: Sampson, Low, Marston, Searle and Rivington, 1893), 64. In Ceylon this simply meant plantation agriculture, but in fact it broke with the tradition of planting coffee under shade trees.

64 *SLNA Despatches* 5/35 4 July 1848 Lord Torrington to Earl Grey.

65 *Ceylon Examiner*, 16 June 1852.

66 On the importance of the notion of rationality as a sign distinguishing European imperialists from other peoples, see Adas, *Machines*.

67 This worry about the unreliability of certain types of imperialists is discussed at length by Stoler with particular reference to the Dutch East Indies in the nineteenth century. See A.

make planting a success were institutions that could collect and communicate locally specific information that could be supplemented by modern scientific knowledge of plantations elsewhere and a class of self-disciplined, modern, professional managers who could more effectively control the workforce.[68] During the early 1850s just such an institutionalisation began to be attempted. Links were strengthened with the Botanical Garden at Peradeniya, which opened a second research centre at Hakgala in the high mountains above Kandy. The Planters' Association was formed in 1854 to represent planter interests to the government and to collect and diffuse

Figure 4.1   Peacock Hill coffee estate

Source: O'Brian's *Views in Ceylon*, 1864.

technical information useful for planting. Increasingly, during these years, agency houses (management companies) managed estates and rationalised business by taking over sales and export and by sending agents to advise planters on the latest agricultural technologies and plantation management strategies. Central to these modernisation efforts was a significant increase in locally produced handbooks, memoirs, newspaper articles, and the *Proceedings of the Planters' Association*. Scientific papers were prepared on everything from weather, soil, plants, and disease to plantation management. This literature was a mix of technical advice and moral judgement, all framed to produce a new, improved rationalised space, tailored to work in this particular environment.

Stoler, 'Rethinking'; Stoler, *Race*; Thomas and Eves, *Bad Colonists*.

68 For a more detailed discussion of the anxiety surrounding self-discipline in nineteenth century Ceylon, see Duncan, 'The Struggle'.

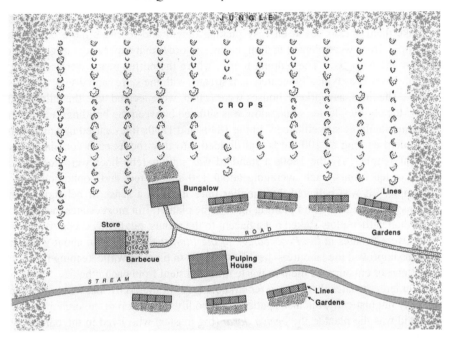

Figure 4.2   Diagram of an estate

Although there are many different species of coffee, there have only been a small number of commercially important ones. Of these, Arabica (*coffea Arabica*) has been dominant. It was the original export coffee from Arabia and is still widely thought to have the best flavour. While Ceylon has an indigenous wild coffee (*coffea Travacorensis*), it has never been cultivated commercially. It was Arabica which had been brought to Ceylon by Muslim traders and subsequently grown by European planters and non-European smallholders alike during the coffee period on the island. Coffee trees are extremely sensitive to climatic and soil conditions. Different growing conditions impart different flavours to the coffee. Not only does the taste vary between places, like wine, the taste of coffee from the same plant varies from one year to the next. Trees take four to five years to produce their first crop and may continue producing for 20 years. One worry for planters was the unevenness with which trees bore fruit. The trees are very sensitive to rain and drought, but can occasionally produce a very large crop inexplicably followed by several years of smaller crops. Some have identified two and seven year cycles, but the combined effect of the sensitivity to weather and natural fluctuations in productivity made it difficult for planters to project yields.[69]

The average plantation covered 100 acres and employed between 100 and 200 workers.[70] Specifications were developed for cutting and burning the forest, and for

---

69 Topik, 'The integration', 25.

70 Snodgrass, *Ceylon*, 22. In the 1840s the rule of thumb was two labourers per acre of coffee, but by the mid-sixties that number was halved due to improved efficiency on the

the staking, holing, and orderly planting of coffee trees.[71] The trees were topped to make them spread out at an easy picking height and to cause more berries to develop.[72] Trees were planted in straight lines in order that supervisors might better observe the workers.[73] The pulping house, where the coffee beans were separated from the fruit, was typically located on a stream, with the stores and drying grounds nearby. The lines, as workers' houses were called, were located near the pulper and stores. Ideally the planter's bungalow was situated between the buildings and fields to further facilitate surveillance.[74] In the 1840s and 1850s lines varied in length, but were often as long as 100 yards and divided into tiny rooms each containing six to eight people.[75] By the 1860s a standard was set whereby lines were composed of six to ten rooms, each averaging about 150 square feet and containing 10 to 20 people.[76] It was believed that smaller lines instilled a sense of belonging and decreased both fighting and plotting against the planter.[77] In most estates there were individual garden plots staked out adjacent to the rooms, but this was controversial. There were debates in the *Proceedings of the Planters' Association* about whether gardens improved the labourers' health at no cost to planters while keeping them on the estates or encouraged them to waste time and steal from each other.[78]

On larger estates there was a remarkable level of hierarchy that was expressed not only linguistically, but also spatially and bodily. At the top of the organizational pyramid was the planter, the *periya dorai* (big master) who lived in the bungalow. Immediately beneath him was his, usually younger, European assistant, the *sinna dorai* (small master), who lived in a smaller bungalow. Below them were 'the native staff', consisting of an accountant, (*kanakapillai*), and others who ran the office. They were usually from the north of Ceylon and lived in what were called 'quarters', not 'bungalows' or 'lines'. The quarters were between the assistant's bungalow and the lines in terms of quality and size. Staff paid the wages, checked the weighing of coffee and rice supplies. Beneath the staff was the head *kangany*, who had overall charge of the labourers. He had large rooms at the end of the lines. Beneath him were the *sesilara kanganies*, (assistant supervisors), in charge of the various labour gangs. They had rooms within the lines that were superior to the labourers, but inferior to

---

plantations and improved transportation off the plantations. This helped alleviate labour shortages in the following two decades. Vanden Driesen, *Indian Plantation Labour*, 79.

71 Lewis, *Coffee*, 37–44; Sabonadiere, *The Coffee Planter*; T.C. Owen, *First Year's Work on a Coffee Plantation: Oonoogalla Estate, Madulkelle (being the essay which received the second prize from the Ceylon Planters' Association in 1877* (Colombo: 1877); Brown, *The Coffee*.

72 Snodgrass, *Ceylon*, 22.

73 Brown, *The Coffee*, 13.

74 Lewis, *Coffee*, 44.

75 Vanden Driesen, *Indian Plantation Labour*, 12.

76 Sabonadiere, *The Coffee Planter*, 65–66; Brown, *The Coffee*, 20.

77 *PPA*, 1876–77, 195.

78 Ibid., Stoler, 'Plantation Politics', 139, argues that plantation gardens in Sumatra were an important part of the plantation system in that they made it cheaper to feed workers. But this also, as the Ceylon planters were well aware, came at the cost of retarding the proletarianisation of the work force.

the head *kangany*. Worker quarters within the lines tended to be by gang, which were in turn organised by caste and village origin.[79]

This hierarchy was, as one would expect, very thoroughly embodied as well. For example, young European assistants tended to treat superintendents with considerable deference and often expressed fear in their presence. However, the greatest divide was racial. Labourers were not allowed to look a planter in the face and were expected to step out of his way if they met him on a path. Furthermore, labourers were never to speak directly to a planter except in response to a question.[80] Men were not to let down their sarong (from its hiked up position for work).[81] Further sumptuary rules forbade umbrellas and other clothing that indicated high status.[82] An unintended consequence of such clear social distance between planters and labourers was the enhancement of the *kangany's* power as virtually all interaction had to pass through him.[83] The linguistic, bodily, and spatial organization of the labour force on the estate was a hybrid mix of European and Indian ordering practices; all were aimed at rationalising spatial arrangements in order to produce a well-ordered labour force. However, as I will show, the consequences in terms of enhanced *kangany* power significantly undermined the authority of the planter.

The Production of the Abstract Body

The Un-cultivated Body

As I have suggested, the bodies of labourers were a principal target of the exercise of power. Various micro-technologies of control were developed in an attempt to transform Tamil villagers into abstract bodies. Labourers, exhausted from their trip from Madras to northern Ceylon, were forced to march a further 150 miles in six days before reaching the estates. Often the *kanganies* misappropriated the head money and provided insufficient food for the trip. Consequently, labourers arrived weak and sick at the estates.[84] The Tamils were figured by the British as base humans, more animal than human in many ways. Dr. Kynsey, the Principal Civil Medical Officer spoke of the labourer, (or 'coolie' in the terminology of the time) as 'naturally lazy, indolent and docile ... He has strongly developed animal

---

79 Kurian, 'Labour', 183–85.

80 All of this was obviously 'in theory', for as we saw in Chapter 3, planters often had sexual relations with labourers. The tensions between feelings of desire and abjection on the part of planters, and sex as a 'weapon of the weak' on the part of labourers was a defining element of the 'sexual economy of colonialism'.

81 The sarong was hiked up for work. To lower it would imply that the interaction with the planter was not work-related and therefore might be social.

82 Kurian, 'Labour', 183.

83 S.B.D. De Silva, *The Political Economy of Underdevelopment* (London: Routledge and Kegan Paul, 1982), 329.

84 K.M. De Silva, *Social Policy and Missionary Organisations in Ceylon 1840 to 1855* (London: Longmans, 1965), 264–65.

passions…'.[85] Some believed that the character of the Tamil workers represented human nature 'in its original, uncultivated state'[86] and therefore there was hope that they could be transformed through the modernising effects of the plantation system. For as Governor Robinson wrote in a despatch to the Colonial Office,[87] '[t]he Tamil coolie is perhaps the simplest, as he is certainly the most capricious, of all Orientals with whom we have to deal in Ceylon. He is like a child requiring the sense of the strong arm of power. He must know that he is subject to parental authority.'[88] As Thomas points out, the other side of European paternalism is the infantilization of native peoples through the imagery of irrationality and inability to survive unaided by Europeans.[89] Having said this, at the time the lower classes in Britain were also often compared to children who were incapable of adult reasoning.[90] The language here is of paternalism and improvement, of capitalism as providing pastoral care, and of authoritarian governmentality. The reality as we shall see, both in this chapter and in Chapter 5 on health, is that there was little benevolence; labourers were in fact worked to what was judged by planters to be the limits of survival. And very often, workers did not survive.

Disciplining the Tamil Body

It was hoped that respect for authority would be instilled through the routines of abstract space. Just as the Tamil workers would cultivate the soil, so their work discipline ought to cultivate their own nature. Careful calculations of tasks assigned, time worked, and pay were based on the relation between the 'flawed but improving' Tamil body and an ideal, abstract body that was based upon European experience and experiments.[91] Such calculations drew upon new forms of bio-medical knowledge

---

85 D. Kynsey, 'Minutes of Evidence, Medical Aid Commission' (*Sessional Papers [SP]*) *Papers Laid Before the Legislative Council of Ceylon* (Colombo: Government Printer 1869), 68.

86 P.D. Millie, *Thirty*, Ch. 15.

87 *SLNA Despatches* 5/53 27 May 1866.

88 As Lambert (*White Creole Culture*, 6, 31–32) points out, by the late eighteenth century on slave plantations in the Caribbean, white supremacism was being regularly tempered with the ideology of paternalism and duty towards slaves. It is questionable to what extent such a view translated into a better life for slaves.

89 Thomas, *Colonialism's Culture*, 132.

90 Wiener, *Reconstructing the Criminal*, 95.

91 Central to governmentality in Britain was knowledge of the body through statistical calculation of the amount of labour various sized and gendered bodies were capable of and what amount of food was necessary to sustain them. W.R Kynsey, the Principal Civil Medical Officer and Inspector General of Hospitals writes in 1876, 'In calculating work it is usual in England to estimate it as so many pounds or tons lifted one foot; the daily work of an ordinary European engaged in manual labour varies as a rule, from 250 to 350 tons lifted one foot, but to perform these amounts continuously without detriment to health, he must be fed on suitable food.

300 tons lifted one foot = an average day's work.

400 do = a hard day's work.

500 do = an extremely hard day's work, which few men could continue to do.'

and were common not only in the workplace, but in the prison as well. Where ever possible, time and space were tightly organised on this basis. There was a strict work schedule on estates. The tom-tom for reveille was beaten at four am. Labourers took their morning meal from five to six am, and then worked from six am to four pm. Adult males were expected to pick two bushels per day,[92] and were not supposed to carry more than 40 pounds for more than 25 miles in 24 hours.[93] Such precise calculations marked what were thought to be the maximum amount of labour that a worker could survive. The lists of tasks, the time necessary and the cost of each job were calculated.[94] Similar formulas were invented for women and children.[95] Wages were paid according to age and gender. In the 1860s men were paid nine pence per day, women seven pence, and children were paid by height.[96]

The labour market was dominated by cash advances rather than regular wages as there were fluctuations in the amount of labour needed on the estates. For labourers in debt before arriving at the estates and more often whose debt increased while in Ceylon, cash advances were a powerful temptation, as they allowed them to buy enough food to keep hunger at bay.[97] Rather than compete for labour during short seasons by offering higher wages, estates competed by offering higher cash advances. Planters saw this strategy as a double edged sword, however. On the one hand, wages remained nearly static during the whole coffee period and the system tied labourers to specific estates, in theory at least. On the other hand, the advance system made planters worry that gangs of labourers would bolt (leave without repaying their debts), or move to another estate as soon as they paid off their advances. As planters were nervous about making advances, they were more likely to do so during peak months of work, such as harvest time, when it was critical that labourers not leave. During other months planters were less likely to make advances or often even pay

---

Such figures, he added, would need to be recalculated for the smaller Tamil body. (*ARC*. 1876, 116c.) In addition, Kynsey calculated that the amount of food needed by bodies depended not only upon the amount of labour expended but also on climate and race. He estimated that a man of average height and weight required 23 ounces of water and 1/100 of his own weight in food daily. This was broken down as follows: an average man at rest required 1.25 grains of nitrogen and 28 grains of carbon per pound of his weight. A woman needed 1/10 less. If employed, a person needed 1/3 more. (*ARC*, 1876, 115c).

92 Brown, *The Coffee*, 42.

93 Labour Law of 1864. This was communicated to planters in the *PPA*, 1864–65.

94 Brown, *The Coffee*, 95.

95 During the 1840s, the ratio of female to male labourers was one in 38. By the 1860s it had increased to one in six as it was felt by the planters that access to more women would make their male labourers more manageable.

96 *SLNA Despatches* 5/54. 6 March 1867. K.V. Jayawardene, *The Rise of the Labour Movement in Ceylon* (Durham: Duke University Press, 1972), 20, estimates that wages for a 10–11 hour day in the 1830s–40s were 4 to 9 pence, and that by the 1860s were 10 pence per day.

97 Behal and Mohapatra, 'Tea and Money', 143; Behal, 'Forms of Labour', 19 note that in Assam during the nineteenth century wages were also often below the subsistence level and that wages were stagnant throughout much of the period.

Figure 4.3   Coffee stores and pulping house

Source: *Illustrated London News* (1872).

wages due, forcing the labourers to borrow money at exorbitant rates from *Chetty* money lenders or *kanganies*; all of which increased the cycle of debt.[98]

Transformed Bodies

Some planters and colonial administrators believed that bio-political techniques had a certain degree of success. As the AGA for Matale stated in his annual report for 1867, 'the coolie after a residence of a few years is a very different being from what he was when he came; instead of being a weak, stupid, ill-fed, badly clad person, he, from being well housed, well cared for, well fed, and employed at regular and sufficient work, returns to his country a strong, intelligent, lusty labourer, to bring back others willing to try their fortune also.'[99] Dr. Van Dort, the Assistant Colonial Surgeon in charge of the Badulla hospital in the planting districts, echoed this when he wrote, it is the 'genius of labour which transforms the labourer.' His countenance, 'mirrors the newly awakened soul, and the consciousness of powers and capabilities, hitherto dormant and unused, stamps the physiognomy with an expression of manliness and intelligence which is never seen in the raw, uncivilised, newly landed coolie.'[100] The planters and government officials in the planting districts argued against their critics in Britain that the estates were not exploitative but positively transformative. I will

---

98 Bandarage, *Colonialism*, 208–09.
99 R.C. Pole, Assistant Government Agent, Matale, *ARC*, 1867, 40.
100   Van Dort, *PPA*, 1869–70, 55.

continue to trace this debate through the issue of disease in Chapter 5. At times the Ceylon Government, as we shall see in the following chapters, was willing to support planters' desires for autonomy on their estates even when they flagrantly disregarded worker welfare; the Colonial Office, however, was much less likely to turn a blind eye.[101]

Resistance in Everyday Life

Covert Resistance

There were many tactics used by workers to resist the planters' attempts to discipline and transform them; most of these were covert, and a few more overt. People without the power to directly challenge those in authority resort to what Scott has termed weapons of the weak.[102] They, therefore, 'poach', to use de Certeau's term, within the spaces of power.[103] The planters' tried to wring as much labour from the workers as possible for as small a wage as possible. The workers countered this by stealing and finding undetectable ways of minimising their labour output.

Workers tried to establish local standards by testing a new master to see how little work they could get away with.[104] They would seek out 'cracks' in the organization of space. Within these early-established frameworks of labour demand and discipline on each plantation, workers developed various tactics for avoiding work. One was to feign sickness, as planters were obliged by law to feed their sick workers.[105] Another was to return to the lines after the six am muster on a pretext of having forgotten something, and then to hide there for the rest of the day.[106] If the planter searched the lines he often found it difficult to see workers hiding in the dark, smoky interiors. The darkness and smoke gave the worker time to scale the partitions between the rooms in a line and return to the fields.[107] But even while in the fields, the workers could use their superior knowledge of the landscape to their advantage. For example, they would hide behind trees and coffee bushes or slip into the jungle at the edge of the plantation for short naps.[108] Tired workers would routinely lighten their loads by cutting open and pouring out a small portion of coffee or rice from heavy bags.[109] They were also known to abandon the bags in the underbrush where enraged planters would sometimes find them rotting months later.[110] During the harvest, one of the more onerous tasks was separating the beans from the coffee fruit. Having to

---

101 Kurian, *State*, 67.

102 Scott, *Weapons of the Weak*.

103 De Certeau, *The Practice*.

104 Sabonadiere, *The Coffee Planter*, ff. 34.

105 This law was put in place after the government discovered instances of sick labourers starving to death.

106 W.A. Swan, *PPA*, 1869–70, 23.

107 Millie, *Thirty*, Ch. 5.

108 Brown, *The Coffee*, 43.

109 Millie, *Thirty*, Ch. 3

110 Ibid., Ch. 5.

continuously feed heavy baskets of fruit into the pulper, workers would periodically slip the belt on the machine in order to catch a few moments rest.[111]

Another tactic was to pretend to fill one's daily quota. The success of such a tactic depended on the level of vigilance of supervisors. For example, workers on James Taylor's estate in the 1850s were expected to dig 25 holes for young coffee bushes per day or lose half a day's pay. Taylor writes, 'they are the greatest cheaters that ever existed, for they must be called back to every hole they make two or three times before it is large enough...they will come asking for tickets for twenty-five holes though they have only done fifteen.'[112] Other ways of quickly filling one's quota was to strip green coffee off the bushes in order to gather ripe coffee more quickly,[113] or to steal coffee from the store at night and bring it in the next day to fill a quota.[114] Workers supplemented their incomes by accepting bribes from bazaar keepers to drop ripe coffee under the bushes, as the bazaar keepers had been given contracts by the estates to glean coffee dropped in the fields.[115] Workers routinely stole coffee beans and rice from the store or from bags during transportation to sell in the bazaars at the edge of the plantations.[116] The theft of coffee however will be covered in more detail in Chapter 6.

All these forms of resistance utilized fragments of space and time. Workers discovered ways to escape the planters' and overseers' panoptical procedures and to discover places where they couldn't be seen; to learn how to take advantage of moments when supervision was lax; to manoeuvre, as de Certeau[117] put it, within an enemy field of vision. A key role in the counter-surveillance was played by household servants in the planter's bungalow and Tamil women in the lines. They would report, for example, if the planter was ill or drunk and consequently staying at home that day, or if he had left the estate to hunt or visit friends.[118] Once it was established that the planter was not coming into the fields, work would slow down or cease. Such flows of information were nearly impossible to control. The planter and the workers played a cat and mouse game with each trying to counter the other's moves. But the game was dangerous for the workers, as their covert resistance, once detected were punished with public humiliation, beatings and loss of food and at times death.[119]

Open Resistance

One of the central goals of authoritarian biopower was to produce obedient subjects. While covert resistance was continual, workers rarely risked open resistance as the

---

111 Ibid., Ch. 6.
112 Taylor, *Papers*, 23 June 1852.
113 Millie, *Thirty*, Ch. 6.
114 Ibid., Ch. 5.
115 Ibid., Ch. 6.
116 *PPA*, 1880–1881, xxii,xxvi.
117 De Certeau, *The Practice*.
118 Millie, *Thirty*, Ch. 34.
119 Millie, *Thirty*, is a wonderful source of detailed information on this cat and mouse game.

potential costs almost inevitably outweighed the benefits.[120] The defacto sovereignty of the planters allowed them to keep the labourers in a state of mere survival with death being considered unremarkable and often unreported at least until the statistics for the population as a whole became alarming to the government. Planters had power to inflict harsh punishment with virtual impunity. Nevertheless, there were various types of open resistance that workers at times adopted. Perhaps the most common form was insubordination, which could take the form of insolence or openly disobeying an order. George Wall, the President of the Planters' Association wrote in the mid 1860s that, 'the coolies are getting a great deal too much of their own way… [and, unless they are controlled]…they will soon be virtually the masters.' He went on to observe that on plantations where workers were given contracts to weed, the 'independence and insolence of coolies was rampant.'[121] In the 1870s Millie wrote that 'the general submissive character of the coolies has very much changed for the worse', He added, 'in a country like Ceylon, uncalled for disrespect or contempt for an employer, even though shown in a very quiet way, should meet with prompt and immediate punishment.'[122] Insubordination was seen as such a serious challenge to the authority of the planter, that it was normally punished by flogging.[123] It is worth noting again that actions as minor as not moving out of a planter's way quickly enough or even looking directly at him were considered insubordination.

Another form of open resistance was using the court system to attempt to collect unpaid wages. Throughout the coffee period, but especially in the 1840s and again in the 1880s when plantations were failing, the late-payment or even non-payment of wages was a recurrent problem. Sir Emerson Tennent reported to the Colonial Office during the crash of 1847 that planters were not paying wages and furthermore were responding to worker demands for wages with 'blows.'[124] The following year the Governor in a Despatch to the Colonial Office wrote that a portion of the crop was lost on certain estates as workers refused to pick until back wages were paid.[125] In theory, the Service Contracts Ordinance #5 of 1841 allowed labourers to sue planters for non-payment of wages,[126] however, as the Police Magistrate at Kandy pointed out, 'the coolies are nearly all ignorant that any law exists for regulating agreements.'[127] Having said that, during the crisis years of 1847–48 there were 2,584 complaints

---

120 Scott, *Weapons of the Weak* makes this same point with reference to Malaysian peasant farmers.

121 G. Wall, *PPA*, 1865–66, 80.

122 Millie, *Thirty*, Ch. 38.

123 *Ferguson's Ceylon Directory* (Colombo: A.M. and J. Ferguson, 1865–66), 127. The other was coffee stealing. Clause 7 of the Labour Ordinance of 1841 provided for 'prompt and severe' punishment for insubordination. On this see, M. Roberts, 'The Master Servant Laws of 1841 and the 1860s and Immigrant Labour in Ceylon.' *Ceylon Journal of Historical and Social Studies* 8 (1965), 26. Flogging for a variety of offences was the norm on Indian tea plantations as well (Behal and Mohapatra, 'Tea and Money', 157).

124 *SLNA Despatches* 5/34 21 April 1847.

125 *SLNA Despatches* 5/35 11 December 1848.

126 Roberts, 'The master servant laws', 24–37.

127 K.M De Silva, *Social Policy.*

for non-payment in the Kandy Courts.[128] Those who had the 'temerity' to take a planter to court were often sorry they had. For planters took it upon themselves to circumvent the laws of the colony and administer rough justice to their labourers. Exacerbating this problem was the fact that many of the Justices of the Peace in the planting districts were themselves planters and were understandably reluctant to prosecute their fellow planters. J.S. Colepepper, the Superintendent of Police for Kandy in the 1840s wrote in a report about tactics used by planters to scare labourers who complained of unpaid wages. He wrote of one labourer who threatened to file charges who was himself arrested on trumped up charges and jailed for nine days. When he returned to the estate and continued asking for his back wages, he was severely flogged and arrived at the police station bleeding from the nose and mouth.[129] No charges were brought against the planter.

The 1841 law was strengthened in 1865 by the passing of Ordinance 11, which allowed workers to terminate employment with two days notice if wages hadn't been paid in a month. The Planters' Association deemed this ordinance undue government interference in the economic realm.[130] Initially the Government of Ceylon and the Colonial Office had been reluctant to 'interfere' with the lives of labourers on laissez-faire grounds. But as the century progressed, state governmentality was progressively strengthened. The first Factory Act in Britain was passed in 1802 to protect the rights of orphaned children in factories. This was seen as consistent with the spirit of laissez-faire on the grounds that orphans were 'helpless.' Subsequent factory acts before mid-century extended the 'helpless' category to all children then to women and young persons. The Colonial Office decided in the 1840s that immigrant labour in Ceylon, though largely adult and male, fit into the 'helpless' category, and hence the state was responsible for them.[131] This was perhaps one of the few instances where infantilization worked to the advantage of the colonized. The first plantation strike was in 1854 when workers successfully struck for one penny extra to match the pay of road labourers.[132] After this, workers occasionally engaged in brief strikes over working conditions. But they were short lived and restricted to single estates.

Runaway Labourers

Workers were tied to their estates by contracts for it was a crime to leave before they expired, unless the contract was paid off. But many apparently found conditions on individual estates so intolerable that they fled. Such a decision was not undertaken lightly, for workers ran the risk of being attacked and robbed by Sinhalese villagers who were anti-Tamil,[133] or being hunted down and jailed by planters and the police.

---

128 Ibid., 247.

129 *C.O.* 54. 235. 21 April 1847 – Tennent to Grey – Enclosure, cited in Vanden Driesen, *Indian Plantation Labour*, 13.

130 *PPA*, 1865–66, 76–78.

131 Vanden Driesen, *Indian Plantation Labour*, 41.

132 C.T. Arasaratnam, 'A Brief History of the Development of Labour Relations in Ceylon', *Ceylon Labour Gazette*, April, 1970.

133 *SLNA Despatches* 5/29 22 December 1842. For a discussion of the persistence of Sinhalese violence against Tamil workers throughout the coffee period, see G.K. Pippet, *A*

In light of this, it is remarkable that a survey in 1862 revealed that 7 per cent of all Tamil immigrants had been arrested for desertion. How many escaped is not known.

There were many reasons for desertion. Perhaps the most common was the accumulation of an unbearable level of debt. As I have said, from the time of their arrival on the estates, workers were in debt to planters for the cost of their passage from India. Furthermore, they were paid so little that most remained in debt throughout their stay. Consequently, they often borrowed heavily from *kanganies* at an average rate of 120 per cent and when necessary they turned to the *chetties*, the Indian bazaar keepers, outside the estates at the rate of 200 per cent interest.[134] Surveys conducted by the planters in the 1870's revealed that the average worker was one to two years' income in debt.[135] It is important to note that such debt was normally accumulated in order to buy food to merely survive. While the planters outwardly deplored this level of debt, it was the wage structure that produced it. Planters were, in fact, ambivalent about debt, realising that on the one hand that it helped tie the workers to the estate, while on the other it transferred authority away from planters to the *kanganies* and bazaar keepers. And yet while planters worried about losing of control to *kanganies*, they contributed to this by paying the labourer wages to the *kanganies* who in turn distributed them to the labourers after deducting their due. This system of payment not only impacted planter control, but allowed *kangany* exploitation of labourers through the partial payment of wages.[136] The planters realised that this system negatively impacted their control, but felt locked into it by the structure of the coastal advance system that created labourer debt to *kanganies*.

Workers sometimes deserted when they were physically mistreated[137]or when a *kangany* from a neighbouring estate lured them away with promises of advances. At times, although this was rare, they even deserted out of fear of the British. For example in 1871, several hundred workers fled the estates because of a rumour that the British planned a blood sacrifice to the gods of a 1000 Tamils to mark the successful completion of the railroad to the planting districts.[138] Earlier that year, Governor Robinson wrote the Colonial Office that, 'coolies are fleeing back in hot haste to their own country on an absurd rumour that war had broken out in China and that they were to be forced into the service of the Government as camp followers and

---

History of the Ceylon Police. Vol. 1 1795–1870 (Colombo: Times of Ceylon,1938), 156; Bastiampillai, B. 'The South Indian Immigrants' Trek to Ceylon in the Mid-Nineteenth Century', *Sri Lankan Journal of Social Science* 7 (1984), 52–53.

134 Moldrich, *Bitter Berry*, 83–84.

135 *PPA*, 1876.

136 Bandarage, *Colonialism*, 209.

137 Conditions improved somewhat on estates after the 1840s as planters became increasingly concerned that bad conditions on estates would discourage labour from migrating. Planters were very aware that word of harsh conditions including physical abuse and non-payment of wages had spread through south Indian villages in the mid-1840s and consequently that migration had dropped from 76,745 to 42,317 between 1844 and 1846. Vanden Driesen, *Indian Plantation Labour*, 54, 19.

138 *Times of Ceylon*, 29 September 1871.

soldiers.'[139] The ultimate form of open resistance was violence, including murder. But this was rarely seen as a viable option as retribution by the British was invariably swift and severe.[140]

Resistance on the part of the workers to the appalling conditions in which they found themselves was made more difficult for a number of reasons. First, their poor health tended to make them submissive to authority. Second, they were surrounded by Sinhalese villagers who were generally hostile to them, and third, their distance from home in the south of India cut them off from support systems. The Labour Ordinances of 1841 and 1865, punished desertion with a term of up to three months of hard labour. So prevalent was desertion, and so effective was the state in aiding the planters to recover workers, that in the words of the Queen's Advocate Richard Morgan, 'the jails are crammed with scores and hundreds of men, women, and children arrested on warrants of desertion.'[141] As prison was found not to sufficiently deter desertion, planters urged the government to allow them officially to flog deserters, but the government refused.[142] There remained a continual tension between the enforcement practices of the planters and government attempts to protect the rights of workers. However, by the end of the coffee period the Committee of the Legislative Councillors ruled that there had been systematic abuse on the part of officials, some of whom were planters themselves, of warrants to recapture deserters.[143] We can see from this that the planters had been at least partially successful in enrolling the state into their practices of discipline.

Strategies of Domination

The workers' tactics of resistance were countered by various strategies on the part of planters. These were premised on planters' notions of the Tamil workers as thoroughly embodied and therefore, lacking in what the British referred to as honour and character.[144] Of particular concern was 'malingering', since Indians were thought by their very nature to avoid labour whenever possible. Those who claimed to be too sick to work were often sent to the fields.[145] Only if they dropped in the fields were they judged to be truly sick. The planter W. A. Swan told the Annual Meeting

---

139　*SLNA Despatches* 5/58. 2 February 1871.

140　In 1866 a planter was murdered by a worker and in the early 1880s three more were, all for being strict disciplinarians. In each case the killer was hung. See, Pippet, *A History*, 231; *ARC*, 1882, 23c.

141　Digby, *Life*.

142　*PPA*, 1864–65, 163.

143　*SP*, 33, 1884.

144　Sabonadiere, *The Coffee Planter*, 86 writes 'untruthfulness comes as naturally to a Tamil as mother's milk. No dependence whatever can be placed upon the statements of the coolies as a class.' As Metcalf, *Ideologies*, 24, points out, the British portrayed Indians as uniquely predisposed to corruption and mendacity, which had the effect of reaffirming their own moral superiority and right to rule. For a discussion of the importance of 'character' in colonialist thinking, see Anderson, *Imagined Communities*, 137; Stoler, *Race*.

145　Swan, *PPA*, 1869–70, 23.

of the Planters' Association how he deterred malingerers,[146] I went ' to muster on Tuesday morning, picked out and sent to police court, one or two men who had not come to work on Monday; the beneficial effect on the following day's turnout has been most marked.' The usual punishment for workers who didn't work hard enough or made mistakes was a beating. Such was the level of planter sovereignty that they rarely questioned whether they were within their rights to do so. When he was a young assistant, James Taylor wrote that the superintendent on his estate 'thumped a coolie for nearly half an hour till he was nearly dead, then threw him off the estate.'[147] J. S. Colepepper, the Superintendent of Police for Kandy wrote in a report on the condition of labourers that a planter had told him how he had hung a labourer by his hands and 'whipped the skin off him' so he couldn't work for three weeks.[148] Beatings were also administered at times to discipline the 'moral life' of workers. For example, the planter R.B. Tytler caned one of his *kanganies* for having an affair. He also caned two women workers whom he suspected of infidelity to their husbands.[149] Such disciplinarity would no doubt have been justified by Tytler in terms of his paternal duty to his employees, to look after their moral development. Not all planters were in agreement on discipline, however. Millie, although arguing a more liberal line, demonstrates just how embodied the workers were thought to be. He urged that planters beat a labourer 'without hurting his body; but…[to] hurt his feelings..yes, hurt the feelings of a coolie! … in this way. All his comrades jeered at him so that he suffered a great humiliation.'[150] Others, such as Sabonadiere, argued that a pay cut was even more effective than a beating for those who earned so little.[151] The *kangany* Carpen acknowledged that both punishments were used. 'An angry master beats us on the back. A quiet master beats us on the belly.'[152] The potential problem, from the planter's point of view, with both of these types of punishments was that the workers were often in such fragile health that the punishments impaired their ability to work effectively. Planters had to consider that the punishment might be more detrimental than the practices it was intended to deter. There can be no question that planters wielded an excessively harsh form of biopower, where damage to bodies was measured against preserving them as an economic resource.

From the beginning, planters argued that what was good for the estates was good for the colony more generally and this was largely accepted by the government. Consequently, they fostered close ties with the colonial government wherever possible, but resisted interference, keeping their space as separate and autonomous as possible. In the early years planters disciplined their workforces as they saw fit. However, increasingly pressures were brought to bear on them, by the Government of Ceylon, the Colonial Office, and the Government of India to feed workers adequately,

---

146 *PPA*, 1869–1870, 24.

147 Taylor, *Papers*, 23 June 1852.

148 *C.O.* 54. 235. 21 April 1847 – Tennent to Grey – Enclosure. Cited in Vanden Driesen, *Indian Plantation Labour*, 12.

149 Digby, *Life*, 49–50. It appears that men and women labourers were punished in similar fashion.

150 Millie, *Thirty*, Ch. 38.

151 Sabonadiere *The Coffee Planter*, 8; see also Swan, *PPA*, 1869–70, 23.

152 Carpen, *The Diary of a Kangany* (Colombo: Privately Published, No Date),88.

to curtail beatings and to challenge what the planters believed were acceptable death rates among workers. All of which is to say that the state sought to limit the near sovereign powers of the planters over life and death on the estates.

In response, planters pressured the Government of Ceylon, through their representative on the Legislative Council, to put in place laws that would close the cracks in their operations which the workers' tactics continually opened up. One such achievement was the Labour Ordinance of 1865, which punished non-work with short prison sentences. However, what all of this surveillance, beating and other forms of punishment demonstrate is that the British did not have 'much capacity to regulate or transform indigenous society and that ... British rule appears as a 'limited raj' – weak, alien, ill informed, often blunderingly ineffectual – and hence an improbable candidate for Gramscian hegemony.'[153]

The Middle Space of the Kanganies

The *kanganies* occupied a middle space between the planters and the workers and showed little solidarity with either.[154] The planter Millie summed up the *kangany*'s position when he wrote, 'if he attended to the master's interest and worked the coolies, he displeased and lost his interest with them. If he humoured the coolies too much, the master was down on him and threatened him with the sack.'[155] From the 1850s on, *kanganies* were professional recruiters who collected labourers from villages in south India. As we have seen, far from feeling solidarity with their fellow Tamils, such was the rapaciousness of many that they appropriated for themselves the part of the head fee intended to feed the labourers on their trip to the estates, thereby turning the march south into a race against starvation, where the weak and sick were abandoned along the roadside to die.[156] Misappropriating two-thirds of the eight shilling head fee, *kanganies* estimated the number that would probably die along the route from disease and hunger and calculated that they could afford to bring more labourers to the island than needed. Governor Ward wrote of this Coast Advance System, that

> it is no exaggeration to say that many hundreds of these poor creatures perish annually from want and disease, in spite of the precautions taken by both the Government and individual planters, who sent over their *kanganies* with sufficient money to provide their labourers with rice for the road; though these advances are notoriously misappropriated or the coolies could not reach the planting districts in the state of extenuation which is almost universal.[157]

The callousness of the *kanganies* was exacerbated by the fact that the *kanganies* were often of higher caste and class than their workers. They would no doubt have looked upon them as having precarious lives and low expectations.

---

153 Arnold, *Colonizing*, 243.
154 Heidemann, *Kanganies in Sri Lanka*.
155 Millie, *Thirty*, Ch. 4.
156 De Silva, *Social Policy*, 264–65.
157 *C.O.* 54. 337. 15 November 1858 – Ward to Lytton, cited in Vanden Driesen, *Indian Plantation Labour*, 51–52.

In spite of this, or perhaps because of it, *kanganies* exercised great control over workers. The *kanganies* acted as crucial intermediaries between the planter and the labourers. As most planters spoke little Tamil, the *kanganies* occupied the powerful position of translators. The planters feared that the *kanganies* and workers would collude against them for example, by accepting small bribes to check workers off as attending muster.[158] The power of *kanganies* to control labour was acknowledged by the payment to them of about 6 cents 'head' or 'pence' money for each worker per day who showed up at work. While the *kanganies* had a real incentive to turn out the labour force, they also had an even greater one to get paid twice by accepting a small bribe for checking off a worker as present.[159] Knowing this, the planters devised various methods of surveillance over the *kanganies* especially at roll-call.

The centre of power for the *kangany* was the separate rooms he had at the end of each line. It was here workers would come for advice, loans, justice, or simply to pay deference.[160] On large estates, where there was more than one *kangany*, the head *kangany* often tried to consolidate his power by offering sub-*kangany* posts to members of his extended family or patronage network.[161] Whilst this strategy was usually effective, there were cases recorded of sub-*kanganies* defrauding head *kanganies*.[162] Likewise, planters were nervous that their *kangany* might desert with his labourers and keep part of the coast advance.[163] Consequently, the Tundu system was put in place, whereby labourers moving to a new estate had to show a piece of paper demonstrating that they were not in debt to their former estate. However, such was the fragmentation of power that *kanganies* soon learned to turn planters' strategies to their own advantage. A *kangany* strategy, which was successful only during times of labour shortages, was as follows: the *kangany* would ask the planter for an advance, and if he did not get it, he would find an estate willing to pay the amount left on the *tundu* to the old estate and also give him an advance. Ostensibly the advance was to be divided amongst the workers in the *kangany's* gang, but it would appear that this rarely happened. If a *kangany* had 20 workers in his gang and owed the planter 600 rupees in unpaid advance money which his workers in turn owed him, that would work out to 30 rupees per worker. If the *kangany* was able to get 750 rupees at another estate, this would leave him with a profit of 150 after paying off the *tundu* of 600 rupees. However, it was the workers who really suffered from this strategy for they were entered in the new estate's account book as owning 750 rupees or 37.50 rupees each to the new planter as opposed to the 30 they owed the old. In this way their debt to the *kangany* built up and there was little they could do to fight it as they usually had no idea how much they owed.[164] *Kanganies* also fought among themselves over deserters. For example, the Police Report from

158 Vanden Driesen, *Indian Plantation Labour*, 51–52.
159 Bandarage, *Colonialism*, 207.
160 Millie, *Thirty*, Ch. 4.
161 *PPA*, 1873–74, 206.
162 Ibid.
163 *PPA*, 1870–71, 13.
164 Chattopadhyaya, *Indians in Sri Lanka*, 50–51.

Badulla district on 29 March 1867 stated that an estate *kangany* was murdered by five other *kanganies* and labourers, during a quarrel about deserters.[165]

The planters understandably had a constant fear of losing ever more control over their *kanganies*. For although they were 'natives', the *kanganies* held powerful positions. They had inserted themselves into the very heart of this particular, culturally hybrid experiment in modern industrial agriculture. The planter R.J. Corbet bemoaned the dependence on *kanganies* saying that it was not 'compatible with the high character that an Englishman should bear that we should go on as we are doing, begging, borrowing, hiring, crimping coolies from each other, according to our respective notions of right or wrong, stooping to acts we should be ashamed of.'[166] Planters, and other representatives of British domination in Ceylon, had to 'make do' or 'make it (modernity) up as they went along.' Some planters dealt with suspected crimping by beating the names of suspected *kanganies* or estate managers out of labourers caught defecting. The following account is provided by the son of the pioneer planter R.B. Tytler:

> My father would take a chair and the coolie would be held face downward with a coolie squatting at his head and another at his feet and the process of squeezing would begin. 'Soloo-(tell me) whack, whack, whack, a pause, 'Soloo' and then more whacks. If a coolie still held out, the process was adjourned to the next day when it was renewed. Two or sometimes three days elapsed before the information was forthcoming. Then the *dorie* (manager) if he was really guilty had a bad time of it.[167]

The remarks of Governor Gregory point to the frustrations felt by many British over the necessary cultural hybridity of colonial space. Estates could not be *tabula rasa*; there were no abstract spaces, no docile bodies:

> [t]he *kangany* system is no doubt a bad one. There is hardly a planter in Ceylon who would not abolish it if he could. But it is impossible to do so. It is custom and anyone connected with the East knows what a barrier that word is to any innovation however palpably beneficial.[168]

The *kangany* was a threat to the planter because he represented the traces of an often harsh village authority, which intruded onto the imagined abstract space of the estate. These traces of older social and spatial practices challenged white authority. But the planters could use this middle space which the *kanganies* occupied in order to shift blame from themselves. For example, the very high death rates of labourers were blamed, as we shall see in Chapter 5, not on the plantation system itself, but on the viciousness of *kanganies*.[169] For the labourers, the *kanganies* were ambivalent figures as well, for although they managed at times to pit the *kanganies* and planters against one another, more often than not they were caught between two sets of exploiters.

---

165 *ARC*, 1867, 208.
166 *PPA*, 1878–79, 28–29.
167 *Times of Ceylon*, 10 March 1921.
168 *SLNA despatches* 5/59. 9 July 1872.
169 *PPA*, 1869–70, 55–115 reprints Dr. Van Dort's charge that estate workers receive inadequate health care, and letters in reply from planters denying the charges.

Conclusion

The coffee estates were spatially demarcated sites of authoritarian biopower within the colony. At their worst they were 'zones of social abandonment'[170] where the mere survival of labourers was the goal of the planter. Within these spaces, planters exercised a defacto sovereignty over their workers. Because workers strongly resisted attempts to transform them into abstract, docile bodies, the planters resorted to surveillance and force, justifying such coercion on the grounds that the workers were degraded, pointing to their acts of resistance as confirmation of their moral failure. To resist the demands of discipline was considered lazy and spineless; to reject the space-time organisation of abstract space was considered dishonest, demonstrating a lack of self-control; escaping from intolerable sometimes even deadly, conditions was called desertion, a criminal breach of contract, and a failure of loyalty. Although Tamil workers were forced to adapt to the highly routinized estate regime, they never fully embodied colonial European modernism. Hegemony was not accomplished; self-discipline made little sense within the network of limited coloniser expectations and the ambitions of the colonized which characterized the forced Tamil insertion into the hybrid and fissured space of the plantations. Authoritarian biopower was practiced, not for the good of the labourers, but for the benefit of the planter and the economic viability of the colony despite any claims to the contrary.

The planters' strategies of power and their desire to record in minute detail all of their observations and theories, their plans for the rationalisation of space and time, have provided me with the bulk of my data; not the voices, but some of the practices of the workers.[171] It is a sad fact that those who have no voice have only their oppressors to speak for them. Through a close reading of government reports, and British planters' pamphlets and manuals warning each other of workers' tactics of resistance, one can begin to develop a record of the spatial and temporal practices that took place in response to the estate owners' attempts to create rationalised space and disciplined, docile bodies. Just as the workers' place was highly constrained within the bureaucratised spaces of the estates, so it was constrained in the historical record. As the workers resisted domination by creating marginal spaces within the carceral space of the estate, so their resistance now occupies the margins of the colonial documents which themselves were part of the socio-technical apparatus of domination. It is only from the margins of the discourses of domination that an historical geography of the oppressed may be written.[172]

---

170   J. Biehl, 'Vita: Life in a Zone of Social Abandonment', *Social Text* 68 (2001): 131–49.

171   See Barnett, 'Sing Along'; McEwan, 'Cutting Power Lines'.

172   On re-reading colonial archives in order to recuperate subaltern voices, see: R. Guja, *Elementary Aspects of Peasant Insurgency in Colonial India* (Delhi: Oxford University Press, 1992); V. Das, 'Subaltern as perspective', in *Subaltern Studies VI: Writings on South Asian History and Society*, edited by R. Guja (New Delhi: Oxford University Press) 310–24; J.S. Duncan, 'Complicity and Resistance in the Colonial Archive: Some Issues of Method and Theory in Historical Geography', *Historical Geography* 27 (1999): 119–128; Clayton, 'Imperial Geographies.'

Chapter 5

The Medical Gaze and the Spaces of Biopower

Power is everywhere; not because it embraces everything, but because it comes from everywhere.

M. Foucault[1]

Introduction

In this chapter I will explore the bio-political responses to disease among immigrant Tamil labourers by the colonial state and the planters. The coffee period lay at the cusp of three major discursive regimes of medical practice, European miasmic and germ theories, and south Asian *ayurvedic* theories. I will examine how disease and climate, as non-human agents in the heterogeneous ecology of highland Ceylon, interacted with these competing discourses of disease in the struggles over authoritarian biopower among the planters, the state, and estate labourers. Expanding colonial governmentality placed the health of labourers as a primary responsibility of the government and of economic institutions such as the estates. The Tamil immigrant labourers were seen as a special population, as the Government of Ceylon had given an undertaking to the Government of India that they would look after the welfare of these Indian subjects. However, labourers were the often unwilling subjects of European medical practice. I examine the spatial techniques of care and surveillance the government employed in attempting to protect them along the roads to the estates and to identify carriers of cholera, which was arguably the most terrifying disease of the colonial period not only for the workers themselves, but for the local populations who resided along the routes. This chapter also focuses on the coffee estates as defacto sovereign spaces where planters practiced a harsh and authoritarian form of biopower. Of particular interest is the issue of sanitation, which was suffused with moral and racialist overtones. The immigrant labourer's body was a terrain that was hotly contested by planters and government. Both sought authority and planters were wary of what they took to be uninvited bio-political incursions into the largely autonomous spaces of the estates. To complicate matters further, the labourers often resisted as best they could these bio-political initiatives. I trace these struggles through the documents of the Planters' Association, the reports of GAs and medical officers, and the debates of the Legislative Council.

---

1 Foucault, *The History of Sexuality: Volume 1*, 93.

The Rise of Medical Topography

Ceylon was a site at which various different regimes of medical practice came together. The first European colonists, the Portuguese and the Dutch, generally had great respect for Sinhalese medicine and although the Sinhalese in turn valued European medicines, the exchange of knowledge had flowed largely from the Sinhalese to the Europeans.[2] One Dutch governor in the eighteenth century argued that the pharmacopoeia of Ceylon was so complete that there was no need to send any medicine from Holland.[3] At first, the British shared this view, seeing Indian medical practitioners as knowledgeable because they understood the local ecologies of disease better and because British theories of the time shared many foundational principles with their South Asian counterparts.[4] For example, Europeans and Indians held in common theories of disease that assumed a causal connection between climate and the body. European theories began to diverge and fragment as they were challenged by competing schools of thought within Europe, however environmentalism in medicine lasted well into the nineteenth century. People's constitutions were thought to reflect the climate in which they lived; hot climates produced hot tempers, for example. As discussed in Chapter 1, according to the Galenic view, climate was a central determinant of health. Other determinants were: food and drink, sleep and waking, air, evacuation and repletion, motion and rest, and the passions and emotions. By controlling these factors, one could in theory protect oneself at least partially from the impact of climate, but one could never escape the determinism of climate.[5] One of the causes of fever was thought to be the decomposition of humours within the body, hence the frequency with which purging was prescribed as a cure.[6] Likewise, at the environmental level, the decomposition of damp vegetation was thought to produce miasmas which were another major source of fever and other diseases.[7] Although eighteenth century Europeans certainly saw tropical climates as unhealthy, by the beginning of the nineteenth century, the tropics came increasingly to be seen as an extremely pathogenic environment.[8] This change in perception may be explained by the fact that mortality in Britain dropped relative to South Asia due to measures such as draining low-lying areas in Britain.

Until the germ theory of disease became widely accepted at the end of the nineteenth century, the dominant model of epidemic disease was contamination rather than contagion. According to the contamination model, rather than diseases having specific causes, certain environmental conditions were thought to produce

---

2   Dutch embassies to the King of Kandy regularly brought European medicines along with pearls, gold and silver. (C.G. Uragoda, *A History of Medicine in Sri Lanka* (Colombo: Middleway Limited, 1987), 69).

3   Ibid., 68.

4   Arnold, *The Problem*, 44.

5   Worboys, 'Germs', 184.

6   Harrison, *Climates*, 13, 38, 44, 48, 56.

7   J. Johnson and J. Martin, *The Influence of Tropical Climates on European Constitutions* (London: S. Highley, 1846), 88.

8   Harrison, *Climates*, 10.

a range of diseases.[9] For example, miasmatic environments were thought to be the cause of sickness and as such it was environments that needed to be cured, not people. Because, under the miasmatic theory of disease, air was the principal carrier of disease, meteorology was a central pillar of medical practice. The environmental approach to disease required researchers to conduct a medical topography, the mapping of diseased environments.[10] Once identified, these pathogenic spaces could either be avoided or sanitized. In his influential mid-nineteenth century treatise on the tropics and disease, Martin[11] argues that the medical topographer:

> should investigate all the circumstances which tend to deteriorate the human race, and to lower its vigour and vitality; all that relates to the external causes of diseases, their propagation, and their prevention; all plans for improving the physical, and through it, the moral condition of the people. He should cultivate more extensively the medical topography of the empire. The natural features and particularities of every locality affect materially the life and health of the inhabitants. Any general system of sanitary inquiry should, therefore, embrace information respecting the surface and elevation of the ground, the stratification and composition of the soil, the supply and quality of the water, the extent of marshes and wet ground, the progress of drainage; the nature and amount of the products of the land; the condition, increase or decrease, and prevalent diseases of the animals maintained thereon; together with periodical reports of the temperature, pressure, humidity, motion, and electricity of the atmosphere. Without a knowledge of these facts it is impossible to draw satisfactory conclusions with respect to the occurrence of epidemic diseases, and variations in the rate of mortality and reproduction.

Bewell remarks that European perceptions of the tropics as pathogenic spaces helped to justify the insertion of European modernity into those spaces. He states that 'the colonization of bodies thus proceeded from and was largely supported by, the medical colonization of physical space.'[12] This mapping of unhealthy places and the systematic collection of statistics underpinned sanitary reforms as a central program of governmentality which, in the words of Prakash, constituted 'a new order of knowledge and power.'[13] The belief in the need for improved sanitation, although based on ideas of environmental causation of disease, had important implications for it placed governmental action at the forefront of disease control. New theories of disease emphasized the need for organized programs of control. This was a blow not only to liberalism more generally, but also to the planters who feared what they saw as onerous costs due to unnecessary governmental interference in the running of their estates. It is important to understand, however, that it was not just the collection

9   Curtin, *Death by Migration*, 51.

10 Bewell, *Romanticism*, 30–31. By the mid-nineteenth century medical topographies were being conducted around the world (Curtin, *Death by Migration*, 43). On the mapping of diseased spaces in Britain and India see P.K. Gilbert, *Mapping the Victorian Social Body* (Albany: State University of New York Press, 2004) and W. Jepson, 'Of Soil, Situation and Salubrity: Medical Topography and Medical Officers in Nineteenth Century British India', *Historical Geography* 32 (2004): 137–55.

11 Martin, *The Influence of Tropical*, 102.

12 Bewell, *Romanticism*, 34.

13 Prakash, 'Body Politic', 196.

of raw numbers, but the calculation of averages, correlations, and regularities and their use in the development of social, medical and moral theory which represented at least the beginnings of a new way of thinking about populations. During the nineteenth century in Britain and France causal chains and statistical laws were explored and became the basis for policymaking, the revision of ideas of free will and determinism, and ultimately the apportioning of moral responsibility. It was all this that characterized 'the new order' that Prakash points to. In his authoritative analysis of the development of statistical thinking during the nineteenth century, Ian Hacking describes how ideas of determinism were challenged by theories of probability and how these stirred up philosophical and political controversies.[14] As I will illustrate using the case the coffee plantations, the British colonies were important sites of the working out of these ideas.

By mid-century, the British began to place more emphasis on contagion and developing multi-factor models of disease. Increasingly, they believed that South Asian medicine was irrational and dangerous. This view was but one instance of a more general critique of folk knowledges current at the time in Europe and elsewhere.[15] Despite their arrogance and confidence in their scientific research methods, Europeans still did not understand the causes of the potentially fatal diseases of the tropics such as cholera, dysentery, smallpox and malaria.

Western medical knowledge in South Asia in the second half of the nineteenth century can best be understood as state medicine, that is, medicine intended to support the populations of native and non-native groups seen as statistical entities in terms of the economic and acclimatizing goals of the colonisers. The two medical controversies I will be discussing in this chapter are situated at an important intersection of medical history, the last decades of miasmatic theories of disease, when climate was seen as an important causal factor, the early decades of the sanitary movement, just before the fine-tuning of this movement in response to developments in bacteriology and the development of modern governmentality in medicine and health. It was within this context that public health developed in Ceylon, by which I mean a health system which was intended to take into account the needs of the local population and migrant labour. Hygiene and medical theories underwent modification according to developing theories of racial difference and increased understanding of tropical environments. Furthermore, while the nineteenth century saw dramatic changes in European medical theories, elements of older humoural and environmental theories continued to be held late in the century and colonial European medicine often lagged behind that of Europe.

The British had a negative view of the forested highlands of Ceylon based very largely upon the high death rates that European troops suffered during repeated invasions of the Kandyan kingdom. The British remained unfamiliar with many of the fauna and flora in the highlands, especially in the early years of the nineteenth century. As it happened, the miasmatic theory of disease dovetailed conveniently with utilitarian beliefs that such deadly wastelands and jungles should be transformed

---

14 I. Hacking, *The Taming of Chance* (Cambridge: Cambridge University Press, 1990).

15 Arnold, *Colonizing*, 51.

by modern European practices and technologies to make them economically productive.[16]

Samuel Baker,[17] the pioneer who opened up the high mountains around Nuwara Eliya to British settlement, posited a connection between European capitalism and the eradication of disease which, even if wrong in the exact attribution of causes, correctly predicted the correlation. He stated:

> The felling and clearing of jungle, which cultivation would render necessary, would tend in great measure to dispel fevers and malaria always produced by a want of free circulation of air. In a jungle covered country like Ceylon, diseases of the most malignant character are harboured in those dense and undisturbed tracts, which year after year reap a pestilential harvest from the thinly scattered population. Cholera, dysentery, fever, and small-pox all appear in their turn, and annually sweep whole villages away.

Europeans blamed the high rates of mortality in the highlands from diseases such as malaria, small-pox and beriberi in large part on the nature of the forested environment.[18] While in fact it was a highly dangerous environment, the effects of these diseases were exacerbated by Europeans' lack of knowledge about and increasing disparagement of *ayurvedic* medicine.[19] While malaria was introduced from India thousands of years before, various other diseases were introduced or reintroduced by the military and economic networks established by Europeans. For example, the Dutch brought syphilis and yaws to the island, while the British conquest brought rinderpest (murrain) from India, which killed half the cattle on the island in 1800 and subsequently reoccurred throughout the island on a regular basis.[20] Similarly, cholera was spread from India in the early nineteenth century.[21] Likewise, smallpox was reintroduced to the highlands by the British in 1819 after a two decade absence from the area.[22]

A complex mixture of medico-environmental discourses could be seen as late as 1870 in the Annual Report of the Colonial Surgeon for the Central Province. Dickman, the Colonial Surgeon, began by discussing the climate, elevation and soils of the highland region in the form of a medical topography. He clearly favoured a miasmatic theory, arguing that fevers and dysentery are produced by vapours, certain soils and heat. He cited the famous exponent of medical topography, Sir Ronald Martin, arguing that thick decomposing vegetation interacting with water cause the soil 'to exhale noxious gases.'[23] Marshes were to be drained as winds blew miasmatic air onto human settlements bringing fever. While earlier, forests were thought to produce malaria and other diseases, Dickman argued that the relationship of vegetation to disease is highly complex because forests can cool the

---

16 Webb, *Tropical*, 50–52.

17 S.W. Baker, *Eight Years in Ceylon* (London: Longmans, Green and Co, 1855), 75.

18 Webb, *Tropical*, 52.

19 On shifting British attitudes to ayurvedic medicine, see Arnold, *Colonizing*.

20 Uragoda, *A History of Medicine*, 55, 71; Webb, *Tropical*, 28.

21 Ibid., 258–59.

22 Webb, *Tropical*, 29.

23 *ARC*, 1870, 378.

air and 'absorb malarial [vapours], and thus preserve the surrounding neighbourhood from disease.'[24] Here we see the beginning of a British concern for the wholesale deforestation associated with coffee plantations in the interior. He concluded his discussion of medical topography thus: '[n]otwithstanding the progress that science has made, we are as yet unable to determine the share that the weather may have in the production of disease.'[25] Dickman had no doubt, however that

> certain winds have ... been found to influence the condition of ulcers and wounds. Certain conditions of atmosphere have been observed to precede epidemics. And in this country, the north wind is always associated with fever and rheumatism. The state of the air in respect of electricity, its abundance or deficiency, is likewise a powerful modifying agent, and, accordingly, we find that calms are greatly dreaded everywhere, particularly in the East; whilst thunderstorms are looked upon as purifying agents. But in respect to the pestilence most dreaded in the East, cholera, it has been observed that electricity neither influences it in one way nor the other.[26]

 He added that elevation also has an impact on disease. Accordingly, a patient on the coast who suffers from fever, ulcers, and 'constitutional debility, derive great benefit from a run up to Kandy. Those suffering from pulmonary afflictions, rheumatisms, liver and bowel complaints, cannot expect any benefit from the change.'[27] He coupled his discussion of climate and disease with the newer belief in the importance of sanitation. As he put it, '[o]f late great attention has been paid to sanitary service so-called. The proper observance of sanitary regulations ... has been productive of immense good, and averted a large amount of misery.'[28]

Indian Labourers and the Rise of State Medicine in Ceylon

In the early years of British rule, hospitals in Ceylon were run by and largely for the military. The British generally felt little responsibility for the health of the non-European population. However, a series of epidemics beginning in the early nineteenth century helped persuade the government that the health of the non-Europeans and Europeans on the island were intertwined. Over time this would lead the government from state medicine[29] to a much broader conception of public health. By the late eighteenth century, smallpox had been defined as a significant problem in Ceylon. When a vaccine was introduced in 1802, the British immediately set out to vaccinate the population of Jaffna, out of fear that the disease would spread to the British troops stationed in the north of the island. Such a programme overtaxed the small British army medical detachment and so the government created the Native Medical Establishment (also called the Civil Medical Department), composed largely

---

24 Ibid., 379.
25 Ibid., 379.
26 Ibid., 380.
27 Ibid., 381.
28 Ibid., 379.
29 On the concept of state medicine see Arnold, *Colonizing*.

of Dutch Burghers to act as medical sub-assistants to vaccinate the population.[30] In the early years of British rule, the government sought to treat contagious diseases that might spread to the European population, as well as leprosy, and lunacy. Beyond that, it was only missionaries who undertook to provide western medicine to the local population. An exception to this was the care of prisoners, which was seen as the moral responsibility of the system of justice.[31] In 1858 the government removed the Civil Medical Department from military control. This separation was a landmark in the development of public health as it extended rudimentary medical care to a larger sector of the population, both urban and rural, through a network of hospitals and dispensaries.[32]

David Arnold[33] makes the point that Western medicine was more than simply a tool of empire; European doctors working in South Asia found it to be an excellent laboratory for the study of variation in disease, as increasingly medicine began to operate through racialized categories. He has argued that, 'colonial rule built up an enormous battery of texts and discursive practices that concerned themselves with the physical being of the colonised.'[34] While such a concern had a humanitarian basis, it can also be understood in terms of more authoritarian governmentality, the economic and enculturating goals of the colonisers. Hygiene and medical theories were specifically tailored to theories of racial difference and tropical environments. One of the central tenets of colonial governmentality was the division of a population into subgroups or populations as statistical entities with each having different characteristics, different norms, and different needs. It was believed that different technologies of power should be brought to bear on these groups depending upon their characteristics. Different calculations were used to decide what rates of sickness and death were unacceptable, implying that some deaths were acceptable. Tamil plantation workers were thought to be sickly by nature and consequently short-lived. They were also seen to mature quickly, as one planter put it, 'in the East ... a youth of nine or ten is mentally as precocious as our boys at double that age.'[35] Thus it was thought that their labour should best be utilized efficiently before they sickened and died. Given the short-term requirements of labour on the plantations, this was not seen as a serious problem. Such beliefs also naturalised the use of child labour and high death rates, thereby reducing the responsibility of the government or employers. Thus, while the identification of group norms allowed the government to intervene in the lives of such groups, it could also serve as a justification for not acting, or in the case of estate labourers for justifying living conditions that were deadly. Such notions could also legitimate the sovereign power of the state to decide who would live and who would die. For, high death rates were generally viewed as more acceptable in a subgroup that was felt to be 'naturally' short-lived. However, there were debates about what exactly the death rate among the labourers was and about what rate could be considered morally acceptable. High death

30 Uragoda, *A History of Medicine*, 82.

31 Ibid., 88.

32 Ibid., 93–94.

33 Arnold, *Colonizing*, 292–93.

34 Ibid., 8.

35 A Planter, *Ceylon in the Fifties and the Eighties*, 12.

rates were bound to attract criticism in Britain which was experiencing improving health conditions, lowering death rates and new developments in public health. The planters, as I argued in the last chapter, acted not so much to support the welfare of their workers but merely 'to make survive.' However, they were sensitive to charges that they shared responsibility for the high death rates among Tamils, as they feared increased costs might be imposed by the government.

During the coffee period, some Colonial Office officials such as the Colonial Secretary Lord Glenelg and the Permanent Under-secretary James Stephen were evangelical Christians who were very much interested in questions of morality. Another influential group who took an interest in the Tamils was the Aboriginal Protection Society. Likewise, the Indian Government, concerned about the dreadful treatment of Indian indentured labourers in places like Mauritius and the West Indies, passed the Indian Act Number 14 of 1839 prohibiting the employment of Indians abroad[36] While emigration to Ceylon was allowed because it was close to India, the planters were very aware that the Indian Government might cut off the labour supply if it was thought that workers were being seriously mistreated.[37] This would have been disastrous for the planters who could not persuade Sinhalese villagers to work at the wages they offered. Pressures for planters to take a more pastoral role by cooperating with the government's policies for the welfare of agricultural workers came from multiple directions from the late 1830s on. The India Act of 1839 created a 'special relationship' between immigrant Tamil labourers and the Government of Ceylon.[38] The latter was charged with the ultimate responsibility for protecting the workers. However, the labourers found themselves under the defacto sovereign control of planters who for the most part ruled their estates, as we have seen, like little kingdoms. It is within this context that we will explore how authoritarian biopower was worked out on the ground, so to speak, in the highlands.

Death on the Roads

Whilst rudimentary health care was not offered to the majority of the inhabitants of Ceylon until the late 1850s, the government had begun to extend its bio-political gaze over immigrant labourers nearly two decades earlier. The reason for this special attention was twofold: first, as we have seen, the labourers were a population politically earmarked for special care, and second, although no figures were available

---

36 D. Northrup, *Indentured Labour in the Age of Imperialism 1834–1922* (Cambridge: Cambridge University Press, 1995).

37 For a discussion of abolitionist and liberal influences on British policy in India, see Metcalf, *Ideologies*. On evangelical influence on labour recruitment see K.M. De Silva, 'Indian Immigration to Ceylon – the First Phase, c. 1840–1855', *Ceylon Journal of Historical and Social Studies* 4 (1961),110; De Silva, *Social Policy*.

38 The Government of India was more a protector in theory than in practice for as Behal and Mohapatra, 'Tea and Money', 158, document, the Government in India did little or nothing to stop the exploitation of migrant labourers on tea estates in Assam. The labour conditions there were much the same as on Ceylon coffee estates.

in the early 1840s, the death rates along the roads were deemed excessive even for non-Europeans.

From the beginning of the coffee period in the 1840s both the planters and the government worried about being blamed for the high rates of sickness and death among labourers. Consequently, in the early 1840s the government, with a financial contribution from the planters, built shelters along the Puttalam-Kandy road in order to give labourers some protection from the elements on their way to and from the plantations.[39] [See Map 4.2]. In 1843 the government also built hospitals in Kandy and Puttalam for the labourers as well as more shelters along the route and stationed government officials and other employees there to rescue labourers who were too weak to continue.[40] This marked the beginning of a medical regime of practice to make visible the routes to the estates by surveying the labourers and putting in place technologies of care. These practices were driven in the early 1840s by local press reports of the high death rates amongst immigrants. Both the planters and the government realising that they were thought to have bio-political responsibilities, sought to shift blame for failure. The planters made their case in the *Proceedings of the Ceylon Agricultural Society* as follows. First, they conceded that during that year the labourers had suffered from so much fever and dysentery, that alarm spread among them and that many left the estates while sick.[41] However, the planters argued that this sickness was environmentally-produced and thus they were not at fault. For, as one planter stated, 'the early part of the year was unusually rainy, and, it is well known, that Malabars suffer from exposure to wind and rain more than from any other cause.'[42] He went on to add that the labourers had made an irrational decision by leaving the estates where they would have received European medical care. Once on the road, 'the Malabars being extremely penurious would deny themselves the most ordinary comforts in order to take their earnings to their homes, and from the general apathy to the suffering of their fellow creatures, nothing can more easily be conceived than the sick unable to keep up with their parties, should be left to perish on the road.'[43] The none-too-subtle implication was that the labourers themselves were at fault and that by venturing out onto the roads, they had become the bio-political responsibility of the Ceylon government.

The planters also played down the problem, suggesting that the estates were the victims of bad press. 'The accounts of the sickness were exaggerated, even by Europeans; and to such an extent did statements mislead the public, that the Planters, as a body, were charged with indifference to the condition of the labourers.'[44] Having said that, the planters were fully aware that they provided very little medicine for labourers and often not enough food. Hence, they decided to provide some

---

39 Vanden Driesen, *Indian Plantation Labour*, 23; *PCAS*, 1842, 36; *PCAS*, 1843, 3.

40 Vanden Driesen, *Indian Plantation Labour*, 25.

41 Fever, dysentery and dropsy while thought at the time to be specific diseases, from the point of view of contemporary science have to be seen as generic terms for a series of illnesses that were little understood and often confused in the nineteenth century. Vanden Driesen, *Indian Plantation Labour*, 15.

42 *PCAS*, 1843, 2.

43 Ibid., 2.

44 Ibid., 2.

rudimentary care for fear that the government might intervene. One wrote in the Proceedings, 'unless some internal arrangement be made upon estates, higher authorities will most assuredly intervene to compel the adoption of some system which shall exonerate the Agricultural interest of seeing the roads choked up with the sick, the dying, and the dead.'[45]

Government officials did not wish to take blame for the death rates either. In 1846, the Governor, Sir Colin Campbell, wrote to the Secretary of State for the Colonies arguing that if the planters did not take care of their workers then a direct assessment might be necessary.[46] In the first heady decade of the planting boom, speculating planters appear to have made an economic calculation about the value of labour and how much it was worth spending to keep workers alive. As long as they were able to procure fresh labour, it was deemed uneconomic to spend much money on medicine or even on feeding them. The survival of the labour force was simply not a high priority.

In spite of the fact that the estates system was productive of ill-health and all too often death,[47] the planters argued vehemently that they looked after labourers responsibly and that the primary problems were to be found along the roads to and from the estates. There was, of course, much truth to the planters' claim that labourers died of disease and starvation along the roads. Labourers often did arrive at the estates ill or dying of malnutrition, dysentery or cholera. What is at issue here is the systematic attempt by the planters to deny any responsibility for the ill health of the labourers under their care.

In response to the planter's charges the government established a Commission in 1846 to look into the problems of sickness and death along the roads. As it turned out, the Commission found blame all around. The Commissioner wrote that, 'many dead coolies ... [lay] unburied along the side of the [Puttalam-Kurungalla] road', and that there were only 'two sheds on this line of road, fit for the accommodation of the great number of coolies daily passing and repassing to and from their country.'[48] The Committee found that the health of labourers was much worse on their way back to India than on the way to the estates. The government used its systems of surveillance to gather reports on the condition of labourers along different stages of the route. Officials in the planting districts, and the roads leading to them, reported findings that were very damaging to the planters' case. One wrote, 'the coolies were generally poor when they arrived, but in a healthy and working condition.' W. Morris, the AGA of Kurunegalla who oversaw 46 miles of road had this to say:

> The Malabar coolies arriving in the central provinces by the Puttalam road are cleanly and healthy in appearance. They are seldom if ever found sick on the road, nor admitted into hospital at Kurunegalla. When returning to their country they are invariably dirty, their clothes old and usually in rags ... the slovenly, ragged appearance of the returning coolies

---

45 Ibid., 4, quoted in De Silva, 'Indian Immigration', 111.

46 *SLNA Despatches* 11 November 1846.

47 I draw this language from Redfield's discussion of the penal colony in French Guiana. (Redfield, 'Foucault in the Tropics', 65.

48 *C.O.* 54. 235. 21 April 1847 – Tennent to Grey. Cited in Vanden Driesen, *Indian Plantation Labour*, 11.

is in strong contrast to the cleanly and healthy looking newcomers, and whilst on the march, the latter are close and compact, the former straggling; the strong and healthy in the van and the weak and sickly in the rear, a gang of 100 sometimes extending a distance of several miles. The sick at first attempt to keep up with their companions, but gradually drop off.[49]

If this was not a sufficient indictment, the Colonial Secretary added:

> Sufficient care has not been exhibited to ensure their comforts on the estates, to erect healthy and suitable dwellings for their shelter, or to provide rice and other necessaries for their support when located at a distance from the bazaars. When attacked by disease they are sometimes driven off to die, instead of being tended by medical advisers or conveyed to hospitals of the Government; their wages instead of being punctually paid are frequently allowed to be months in arrears or even altogether withheld, and their importunities or irritations silenced by blows or personal restraint.[50]

Finally, The Colonial Secretary played the 'slavery card' in citing evidence in his report supplied by police officials. In reply to a question on the condition of the labourers, Colepepper, the Superintendent of Police of Kandy replied :

> this class of the population is in a much worse condition in the Central Province than the negro slave was described to be in the West Indies in former days, as to their wants, their treatment from their employers, the mode of enforcing discipline and labour, and the remuneration received.

The Police Magistrate of Gampola wrote in reply to the same question, 'conditions in Ceylon were nearly as intolerable as Egyptian bondage or West Indian slavery.'[51]

Conditions on the plantations were so terrible that by the late 1840s labourers were reluctant to come to Ceylon except when driven to do so by hunger in Madras. As economic conditions began to improve in India, planters feared that an insufficient supply of labourers would come and so began to make some slight improvements on their estates. It was difficult to know the exact death rates during the 1840s, because mortality figures were not systematically collected. In 1849 the *Ceylon Observer* estimated that between 1841 and 1848, 70,000 workers, or 25 per cent of those who came to the island died.[52] Ten years later in 1859, A.M. Ferguson, who normally spoke for the planter's interests wrote in the *Ceylon Plantation Gazetteer* that 25 per cent was an accurate figure for the 1840s.[53] To put such figures in context, the death

---

49 *C.O.* 54. 21 April 1847 – Tennent to Grey, Enclosure. Cited in Vanden Driesen, *Indian Plantation Labour*, 11.

50 Vanden Driesen, *Indian Plantation Labour*, 12.

51 *C.O.* 54. 235. 21 April 1847 –Tennent to Grey; evidence supplied by Colepepper and de Saram, cited in Vanden Driesen, *Indian Plantation Labour*, 19. An interesting defense against charges of abuse was employed by slave-owners in Barbados who claimed that slavery was preferable to freedom in Africa. See Lambert, *White Creole Culture*, 1.

52 *The Ceylon Observer*, 4 October 1849.

53 *Ceylon Plantation Gazetteer* (Colombo: 1859).

rate during the worst years on the notorious Devil's Island penal colony in French Guiana was also 25 per cent.[54]

The planter's policy of merely 'making survive' was high risk for them. During this time disease could easily sweep away a malnourished population. The 25 per cent death rate of the 1840s strongly suggested that the planters were not even living up to the low standards of bare life. Clearly the Ceylon Government thought that the planters had exceeded the norms of what an acceptable death rate should be, although no specific number was quoted as to what such a rate might be. Consequently, the Government and particularly the Colonial Office came to believe that planters had abrogated their bio-political responsibility to the population under their control. The result was pressure brought to bear on the planters to improve food rations and sanitation on estates. Consequently, in the 1850s a monthly rice ration was given and labourers were less likely to starve.[55] Although sanitation continued to be deplorable on estates and there was much disease, labourers generally left the estates in better condition than they had earlier. However, the death rate in the 1850s remained around 25 per cent. Ferguson, who wrote for the planter-controlled press argued with justification that the pattern of death had changed and it was now those arriving who were dying on the roads rather than those leaving.[56] This new pattern of death shifted responsibility from the planters to the government as many of the sick and dying labourers had not yet reached the jurisdiction of the planters. The new Coast Advance system, referred to in Chapter 4, where *kanganies* were paid to bring over labourers and feed them was identified as the culprit. In a Dispatch to the Colonial Office, Governor Ward wrote:

> ... gangs of coolies ... are decimated by sickness and fatigue during the long journey from Manaar to the Central Province ... The coolies do not get the benefit of one third of the advance charged to the employer. Hundreds die of actual starvation upon the road.[57]

The abuses of the *kanganies* were greatly exacerbated during the first half of the 1850s when the facilities for labourers travelling on the northern road became even worse than they had been during the 1840s. Those earlier structures had fallen into disrepair due to hard financial times in the late 1840s. To compound matters, *kanganies* often appropriated the sheds and rented them to travellers.[58] Many immigrants arrived in the plantation districts so weak that they were unable to work and were forced to

---

54 Redfield, 'Foucault in the Tropics', 71.

55 Wickremeratne estimates that during the 1860s and 1870s Tamil labourers received more than double the per capita amount of rice yearly than did poor Kandyan peasants. However, the peasants were able to supplement their rice with produce from their chena lands and therefore in many instances may have had better diets. See L.A. Wickremeratne, 'Grain Consumption and Famine Conditions in Late Nineteenth Century Ceylon', *The Ceylon Journal of Historical and Social Studies* NS 3 (1973), 32–34.

56 A.M. Ferguson, *Ceylon. Summary of Useful Information* (Colombo: A.M. and J. Ferguson, 1859), 167.

57 *C.O.* 54. 337. 27 May 1858 – Ward to Stanley, in Vanden Driesen, *Indian Plantation Labour*, 52.

58 Bastiampillai, 'The South Indian', 53.

beg, until they collapsed and were taken to hospital to die. This change in the pattern of mortality removed much of the responsibility from the planters, and put pressure on the government to improve conditions along the roads.

By the late 1850s the Governor, Sir Henry Ward, issued an order to repair and extend the number of resting places, sink wells to provide pure drinking water and appoint medical inspectors for the route.[59] In 1859, cognizant of the bio-political problems posed by the Coast Advance system, he agreed to have the Government pay for a ship, the Manchester, to bring labourers directly from the coast of India to Colombo, thereby cutting out the northern routes and the abuses of the recruitment system. This was a period of tension between a general laissez-faire attitude in government, a desire to facilitate the economic viability of the estates which were still seen by many as the *raison d'etre* of the colony, and the development of humanitarian governmentality in Britain and the colonies. The entry of government into subsidizing the recruitment of labour collapsed in 1861 on economic grounds and planters returned to the Coast Advance system while immigration continued along the northern route. Upon his arrival, the new Governor MacCarthy recognized the problem with the Coast Advance system, but nevertheless decided to disengage the government from the planting industry on laissez-faire grounds. He believed that the Coast Advance system, bad as it was, was 'part of free action by private enterprise'[60] Although the planters were great exponents of laissez-faire whenever government bio-political programs cut into their profits, they were very critical of Governor MacCarthy's cost-cutting measures when governmental support for estates was needed. In spite of Governor MacCarthy's attitude, the laissez-faire policies of the Colonial Office began to crumble during the 1860s. As Vanden Driesen says, '[t]he limits of permissible State action in the cause of welfare, already wide, were now made broad enough to include spheres of economic activity previously regarded as sacrosanct.'[61]

In 1862 and 1863 new rest sheds were built along the northern route. Eighteen wells were sunk to provide clean drinking water and three new hospitals were built. But the government was still unhappy with the basic system of immigration, feeling that by merely providing health care along the route they were avoiding the root of the health problem, the Coast Advance system itself. Consequently in the mid-1860s, the government subsidized the cost of a steamer to bring labourers to Colombo. But this system also failed in 1867 on financial grounds as the majority of labourers opted for the less expensive north road. Stymied once more, the government again tried to improve the northern route. Hence by 1872 there was a network of 19 rest houses, 15 miles apart, and small hospitals supervised by Tamil medical practitioners at regular intervals along the northern route. Each hospital had a staff of servants who travelled the route between hospitals several times per week looking for labourers

---

59 T.C.W. Murdoch to Sir F. Rogers. Emigration Commissioner's Report, copied in *PPA*, 1869–70, 66.

60 *C.O.* 54. 363. 30 October 1861 – MacCarthy to Newcastle. Quoted in Vanden Driesen, *Indian Plantation Labour*, 103.

61 Vanden Driesen, *Indian Plantation Labour*, 109.

in distress.[62] In spite of all this, in 1870, the Principal Civil Medical Officer reported that the mortality of migrants was double that of other races and that many died on the roads from disease and starvation. He added that the mortality in hospitals of Indian migrant labourers found along the roads was 24.5 per cent while that of estate labourers was 20.5 per cent and that of the sick, non-migrant poor was eight per cent.[63] Such differentials strongly suggest that it wasn't just the environment that was unfavourable to survival. In view of reports like this, the Colonial Office felt that Governor Gregory was too much under the influence of planters when it came to the protection of the health of immigrant labourers. The Colonial Office noted that it was particularly concerned about reports of starvation on the North Road and suggested that the government of Ceylon provide free rice rations to immigrants. Governor Gregory successfully argued against this on the grounds that during famine periods in south India, Ceylon would become 'South India's pauper asylum.'[64]

Although the planters were keen to encourage government assistance in matters of health along the roads, they resisted all forms of intervention on the estates themselves. In the next two sections of this chapter, I examine the spatial strategies adopted by the government to stop immigrant labourers from spreading cholera to villagers and the tactics used by labourers to evade government surveillance. Next I focus on the debate over sanitation and health care on the estates and the negotiation of boundaries of responsibility for and complicity in labourer illness. In both cases it is the working out on the ground of governmentality as applied to what were thought of as the abject bodies of labourers that is of greatest interest. Various ideas and their practical application were fought over by the government officials, planters, villagers and labourers within an assemblage of highly unequal power relationships.

Spatial Strategies: Cholera and the Labourer's Body

While cholera can be readily cured today, during the nineteenth century it had a terrifyingly high death rate of over 60 per cent and no known cure. It was a horrifying disease, especially in the nature of the death. A person who contracted cholera was normally dead within three to twelve hours of first showing symptoms. It is worth quoting at length a modern day account of the effects of the disease on the body:

---

62 However, it was not until the mid-1870s that a rudimentary system was put in place for the Sinhalese villagers. Governor Gregory in 1873 introduced a network of dispensaries with outpatient facilities for villagers whom he said were 'still at the mercy of ignorant quacks and devil dancers.' *Addresses Delivered in the Legislative Council* Vol. 2. (Colombo: Government Printer, 1900), 314. It is tempting to conclude that this tremendous differential between the care of labourers on the roads and Sinhalese in villages was affected not only by the much higher death rate of the former and special agreement with India, but because these labourers were of central importance to the coffee industry, whereas the Sinhalese villagers were seen as having little economic importance.

63 *ARC*, 1870, 367.

64 B. Bastiampillai, *The Administration of Sir William Gregory, Governor of Ceylon 1872–1877. The Ceylon Historical Journal*, 12. (Dehiwala: Tisara Prakasakayo, 1968), 20–21.

The toxin produced by *Vibrio cholerae* bacteria paralyzes the gut in such a way that intestinal cells secrete water and electrolytes, resulting in diarrhoea and extremely rapid dehydration. Persons who are acutely symptomatic suddenly begin to expel an unbelievable volume of diarrhoea and vomit—10 per cent of a person's body weight can be lost in a matter of hours. The stench of diarrhoea and vomit become overwhelming. The rapid dehydration leaves cholera patients weak and thirsty, their arms and legs grow cold and clammy, and powerful cramps seem to shrivel their limbs and tie them in knots. The tips of their tongues and their lips turn blue, their eyes sink back into their sockets, and their skin hangs limply on their bodies. A fifteen-year-old can be mistaken for a person of seventy.[65]

We now know that cholera bacteria are transferred through food, water or bodily fluids. They can live for days in water and survive in the human gut for weeks without symptoms. But an asymptomatic person excretes cholera-infected faeces. Hence, even a small number of asymptomatic persons can keep the disease alive and spreading, seemingly at random. As cholera spread silently along land routes and through water supplies, the unpredictability of the disease was terrifying. Healthy people are usually able to secrete substances in their stomachs that help protect against cholera, but those who are starving, or even physically run-down, secrete less of these substances and are much more susceptible to the bacillus.[66] This was a major reason why cholera was much more devastating during famine years in South Asia.

The cause of cholera was not fully understood until the mid-1880s. Throughout most of the nineteenth century, cholera was understood in terms of Galenic and other climatic theories. Although a breakthrough in diagnosing the cause of the disease came in 1854 when John Snow traced a cholera outbreak in London to a particular pump, his findings were disputed by many in Britain and South Asia. Until the late nineteenth century, it was common for doctors to associate water quality generally with health, but not to see a causal connection with particular diseases.[67] While some European doctors in India accepted Snow's water-borne theory as early as the 1860s, key figures in the South Asian colonial medical establishments held that cholera was not water-borne but a place-based, miasmatic disease transmitted through the air. They continued to hold this view until Koch's studies in Calcutta in the 1880s removed all doubt that it was in fact water-borne.[68] Some even held onto it beyond this time. What Arnold and Harrison have pointed out in regard to India, which applies equally to Ceylon, is the fact that the water-borne theory challenged long-held environmentalist theories of the uniqueness of the South Asian environment and that this had important economic implications for the state. The air-borne theory

65 C.L. Briggs and C. Mantini-Briggs, *Stories in the Time of Cholera: Racial Profiling During a Medical Nightmare* (Berkeley: University of California Press, 2003), 1.

66 S.Watts, *Epidemics and History: Disease, Power and Imperialism* (New Haven: Yale University Press, 1997), 170–72.

67 Curtin, *Death By Migration*, 54.

68 M. Harrison, 'A Question of Locality: the Identity of Cholera in British India, 1860–1890', in *Warm Climates and Western Medicine: The Emergence of Tropical Medicine, 1500–1900*, edited by D. Arnold (Amsterdam: Rodopi, 1996), 133–59.

allowed the state to adopt a much more laissez-faire approach, while the water-borne view called for a much higher level of governmentality in the form of large-scale sanitary engineering works.[69] Furthermore, the Indian and Ceylonese Medical Services had become rather stagnant institutions where innovation and new ideas, medical and otherwise, were largely discouraged.[70]

Until the early nineteenth century, cholera was a regional disease of the north east of India. But in 1817 an epidemic broke out in Bengal and was spread by the movements of the British army and its camp followers to other parts of India. From being a regional disease, it rapidly diffused around the world during the nineteenth century. As Evans points out, cholera became a quintessential disease of the industrial age, spreading through increasingly rapid and interconnected networks of transportation and through densely packed cities with poor sanitation.[71] There were six cholera pandemics in the nineteenth century.[72] Cholera reached continental Europe during the second pandemic, arriving in Britain in 1831. Evans argues that cholera was 'horrifying and deeply disgusting in an age which, more than any other, sought to conceal bodily functions from itself ...'[73]

The epidemic in Britain focused attention on the poor as conduits of disease from India and intensified class anxieties leading up to the Reform Bill of 1832. The discovery of this horrifying disease among the urban poor in Britain and the poor in India meant that: 'both populations simultaneously emerged as "health threats," the framing of each refracted through the other.'[74] In Britain alone, 130,000 people died of cholera in the nineteenth century.[75] But this was nothing in comparison to India, where it is estimated that 15 million died between the first epidemic in Bengal in 1817 and 1865, and another 23 million died between then and 1947. The Madras Presidency, the area where most of the estate labourers for Ceylon came from, was one of the most affected by cholera. In the epidemic years of 1876 and 77 the mortality rate was staggeringly high. In 1876, 148,193 died of cholera in Madras, and the following year, which was also a famine year and consequently people were more susceptible, 357,430 died.[76] The cholera pandemics of the nineteenth century elicited an unprecedented international response which included ten international sanitary conferences set up to frame measures to stop the spread of the disease. These included periodic quarantine measures at Suez and in Europe against Indian shipping. The Government of India and London vigorously opposed these measures

---

69 Arnold, *Colonizing*; Harrison, 'A question of locality'.

70 Harrison, *Public Health*, 49, 53.

71 R.J. Evans, 'Epidemics and Revolutions: Cholera in Nineteenth Century Europe', in *Epidemics and Ideas: Essays on the Historical Perception of Pestilence*, edited by T. Ranger and P. Slack (Cambridge: Cambridge University Press, 1992), 149–74.

72 1817–23; 1826–37; 1841–59; 1863–75; 1881–96; 1899–1923.

73 R.J. Evans, *Death in Hamburg: Society and Politics in the Cholera Years, 1830–1910* (Oxford: Clarendon, 1987).

74 Bewell, *Romanticism*, 253, 270.

75 Watts, *Epidemic*, 167.

76 Arnold, *Colonizing*, 161, 164, 167. During the famine in India of 1876–78 it is estimated that there were 2.6 million deaths in Madras alone. For a discussion of the convergence of natural and human factors in the famine see Davis, *Late Victorian*, 111.

on the grounds that they believed cholera to be air-borne and did not wish to expand the public health responsibility of government. Nevertheless, cholera became central to debates about the role of government in public health in India.[77]

By the time cholera reached Ceylon from the south of India for the first time at the end of 1818, southern India was already seen as a reservoir of epidemic diseases. Arnold[78] describes cholera as a 'highly political disease' in that it was seen as alien and the failure of the British to bring it under control threatened British power. During the coffee period beginning in the 1840s cholera struck regularly in Ceylon due to the annual migrations of workers from south India. Between 1842 and 1878, around 72,999 people died of the disease on the island.[79] Although continuous records are only available from 1842, it is clear that the disease appeared virtually every year on the island to a greater or lesser extent. During the 1840s an average of 1,073 people per year died on the island of the disease. But their number varied wildly, from 3,881 in 1846 to 19 in 1848. During the 1850s the annual average more than doubled to 2,431 per year with a high of 7,936 in 1854 and a low of 22 in 1856. The 1860s followed the same pattern with an average death toll of 1,972 per year and highs of 5,926 in 1866 and 6,904 in 1867 and a low of 0 in 1868. And finally in the 1870s, the period I will be examining in more depth, the annual death rate was 2,025 with a low of nine in 1872 and a high of 11,963 in 1877.[80]

The Medical Report for 1870, written by the Colonial Surgeon for the Central Province, Dickman, outlined what he observed about the causes of cholera in Ceylon. He stated that, 'most frequently the disease follows in the wake of immigrants from the southern coast of India, although occasionally it shews itself in a sporadic form in different localities, without any appreciable cause, leaving Medical men as ignorant of its causation as they were before.'[81] While he stated that he did not wish to enter into the 'vexed question, whether cholera is contagious or epidemic ...' he said that it is 'transportable, and its communicability is mainly effected through the means of human agency, outbreaks of the disease being generally preceded by the arrival of persons from tainted districts ...'.[82] He went on to say that he accepted the *Report of the Cholera Epidemic of 1866 in England* and that following the work of Dr. Snow there is no question that 'impure water powerfully influences the spread of cholera ...' He considered cholera to be 'an alkaloid organic poison, soluble in water ... and that one cholera patient may disseminate many millions of millions of cholera particles in water ...'.[83] However, not all doctors in Ceylon were willing to abandon a belief in atmospheric influences. For example, his superior W.P Charsley, the Principal Civil Medical Officer, in his 1871 report argued that:

---

77 Harrison, 'A question of locality', 133–34; Arnold *Colonizing*, 191.

78 Arnold *Colonizing*, 159.

79 *ARC*, 1878, 165c–174c.

80 These mortality figures are complied from the Medical Report on Cholera in the *ARC* 1878, 168–69c.

81 *ARC*, 1870, 386.

82 Ibid.

83 Ibid., 387.

both cholera and small-pox appear to require some condition of the atmosphere, which is still unascertained, and which may, or may not, be assisted by local conditions and influences to assume an epidemic form; but experience has taught us that when they do become epidemic, those localities suffer the most in which sanitary laws have been most neglected.

To hammer home the point, he concluded, 'filth and dirty water, poisoned by unwholesome accumulations and stagnation, do not alone create cholera and small-pox ... they appear to require a certain favouring condition of the atmosphere, the exact nature of which still remains to be discovered.'[84] In the Principal Civil Medical Officer's statement, we can see the attempt to straddle and reconcile two different discourses of disease.

It is a mark of the supreme importance of the estates to the government that even during years when cholera was widespread in south India such as 1875-76, immigration to the island was not halted. There was also pressure from the Government of Madras, which lobbied the Government of Ceylon during those years, to allow immigration to continue in order to provide relief for a portion of its sick and starving population.[85] However, demands from the coffee estates for labour had to be balanced with programmes for minimising the spread of cholera. The island's Principal Civil Medical Officer admitted as much when defending the fairly weak measures he had put in place to weed out sick labourers entering the island; he wrote apologetically that he had tried as little as possible to interfere with the importation of labour.[86] It is nothing short of astonishing that in a 'cholera year' in south India such as 1876, 165,000 labourers should have been admitted to work on Ceylon estates. For it was known from experience that such a decision would result in Sinhalese and Ceylonese Tamil villagers dying from coming into contact with infected immigrants. For example, the Colonial Surgeon for the Central and North-Central Province reported that in 1876 outbreaks of the disease 'occurred chiefly in villages near the road ... and that the inhabitants of some Sinhalese villages near Anuradhapura suffered severely.'[87] The sovereign power of the state to make decisions that resulted in deaths among members of the population was demonstrated in the fact that officials weighed the economic costs and benefits of migrant labour against humanitarian concerns and came up with an acceptable death rate. There was a strong geographical dimension to this ultimate exercise of biopower for the choice of immigration routes entailed decisions about which were the regions where people would die and which were the regions where people would live.

The Principal Civil Medical Officer and Inspector General of Hospitals posed what was widely seen as the great dilemma of estate coffee production when he wrote in 1876,

intimately connected as the subject of the immigration of coolies is with the great coffee industry and the material progress of the Colony, it is equally connected with, and I am

---

84 *ARC*, 1871, 462–63.

85 Bastiampillai, *The Administration*, 23.

86 *ARC*, 1876, 123c.

87 *Report of the Colonial Surgeon, Central and North-Central Province*, ibid., 128c.

sorry to state answerable for, the introduction and dissemination of no small amount of such infectious diseases as cholera and smallpox throughout the island ... .[88]

He continued, the germs are 'invariably imported by Malabar immigrants' and 'wherever the Indian cooly travels, he carries with him the elements for propagating these diseases.'[89] He went on to point out that cholera was only found in districts visited by immigrants. The government's task was to figure out how best to exclude those who were visibly ill, insulate the native population from the immigrants, and treat those immigrants who fell ill while on the island. One can identify two spatial strategies of surveillance that the colonial authorities adopted to block the entry of diseased bodies into Ceylon. The first I term the 'choke point', a location where the diseased were identified and quarantined. The second strategy I call 'the quarantine route', a carefully demarcated route where those who managed to escape detection at the choke point could be surveyed periodically as they continued on their journey to the estates.

The Choke Point

Health inspectors were posted at ports in order to detain and treat the sick. But this was relatively ineffective because inspectors found it difficult to diagnose the early stages of cholera and labourers resisted detection. As W.R. Kynsey wrote in frustration,

> Unfortunately, in the absence of all knowledge of the agents which produce cholera or of the articles to which the germ, poison or whatever it may be called, can become attached, and of the reluctance on the part of the Tamil cooly to afford information for fear of detection, as to the state of his health, and of the well-known untruthfulness of his answers to all enquiries, when it is his object to conceal disease. I consider as regards the detection of cholera, the most careful medical inspection at Tuticoreen [in South India] would be a farce, and the same remarks apply with equal force to medical inspections at Colombo. No doubt, smallpox and other eruptive fevers can be detected and cholera when developed, but the latter disease certainly not in its early remediable and probably most dangerous stage for the propagation of the malady, if the patient decides to conceal it.[90]

Dr Kynsey laid much of the blame for the failure of the chokepoint strategy on the ignorance and moral failure of immigrants who sabotaged the strategy. The fact that sick immigrants were not surveyed on board the boats made it easier for them to remain undiscovered, as Kynsey stated:

---

88 Ibid., 122c.

89 Ibid., 122c.

90 Ibid., 124c. Although Kynsey refers to 'germ', the germ theory of disease was not theorised until two years later when Louis Pasteur presented his findings to the French Academy of Sciences in 1878. (See L. Pasteur, 'Germ Theory and its Application to Medicine', *Comptes Rendus de l'Académie des Sciences* 86 (1878) and L. Pasteur, 'Extensions of Germ Theory', *Comptes Rendus de l'Académie des Sciences* 88 (1880).

it is notorious the efforts made by passengers and masters of vessels to frustrate medical inspection and to avoid the expense, trouble and annoyance of quarantine. During the last twelve months many cases have been brought to notice of coolies being dressed as traders, and the bodies of cholera patients being thrown overboard from native crafts during the voyage and no report made on arrival here; their bodies subsequently being washed ashore at Colombo, and after a careful post-mortem examination, the disease they died from was identified as genuine cholera - both these frauds were perpetrated to avoid quarantine, the persons responsible for them little caring what amount of death or misery resulted from the introduction of disease into a densely packed city or province.

Here the failures of modern science to diagnose the disease were compounded by what was deemed ignorance and moral failure on the part of the immigrants who failed to discipline themselves in the interest of their own bodies or the body politic. From the point of view of the immigrants, medical surveillance was yet another regime of oppressive power to be evaded if possible. This was particularly so in the case of cholera which was seen as punishment from the goddess Amal. She was thought to be more powerful than the British and attempts to cure the disease would only anger her, making the disease worse.[91]

The second component of the choke point strategy was hospitalisation of the sick and quarantine of those in intimate contact with them. But this was highly problematic as well. If an infected person was found on a ship, they were removed and the passengers were observed on shipboard for 48 hours. The ship was then 'purified and disinfected ... from germs of disease.'[92] While it was held to be desirable to quarantine the ship for a longer period, this was hardly ever possible because of the 'filthy conditions of vessels often heavily laden with animals, merchandise, food and human beings.'[93] Once again, it was held that the moral failure of the natives to keep clean undermined rational European methodologies. Land quarantines were also established. In 1870 a cholera hut to be used for quarantine was constructed on a hill outside Kandy where labourers entered the city from the north road.[94] In 1876 a temporary quarantine 'cooly depot' was established beyond the Kelani railway station, outside Colombo, where infected gangs of labourers were transported by train and kept under surveillance until the danger of the disease spreading was thought to have passed. But these were seen as only partially effective methods. Dr. Kynsey, citing a report of the Great Vienna Sanitary Conference of 1874, argued that such land-based quarantines were of questionable effectiveness as there was no certain way of containing disease.[95] The choke point strategy thus was thought to be of limited utility in excluding cholera and began to be used as a subsidiary strategy to the quarantine route.

91 Witness before the Cholera Commission in 1867. *SP*, 1867, 20.
92 *ARC*, 1876, 123c.
93 Ibid., 123c
94 Ibid., 386.
95 Ibid., 123–24c.

The Quarantine Route

Throughout the whole coffee period the majority of the Tamil labourers left India at Paumben and Divipatam and took what became known as the great northern route, entering the island at Pesalai, and Mannar and proceeding down the central road through Mihintale, Dambulla and Matale and then on to Kandy and the estates. A smaller number travelled south along the coast road to Kurunegala via Puttalam. In 1876, 104,312 immigrants arrived via these routes.[96] The advantage of these routes over the Colombo route, from the point of view of Dr. Kynsey, was that the infected immigrants who had passed the choke point inspection undetected had to walk through long stretches of very sparsely populated country and therefore were less likely to spread disease than if they moved through more densely populated portions of the island. While these routes had initially been chosen by immigrants as the shortest and cheapest way of reaching the estates, officials were quick to see the health advantages to the settled population. Officials thus stumbled upon a spatial strategy that operated within the parameters set by the incubation period of the disease. The incubation period was typically only two days. After this period someone who contacted the disease would normally show symptoms.[97] Immigrants arriving in Mannar at the start of the North Road would face a six or seven day walk along the North and Central Roads before reaching the estates. By then, the majority of cases would have developed. We will shortly examine in more detail how the quarantine route was developed around the parameters set by the incubation period of the disease. It nicely illustrates Foucault's point that bureaucratic regimes of practice, in this case public health, call into being different forms of expertise and historical assemblages of practices. It also illustrates how certain forms of practice presuppose certain identities, in this case, the irrational, deceitful, ignorance-laden peasant. However, this example of authoritarian biopower did not come without a cost to the population. It entailed decisions about which parts of the population would be exposed to cholera and which would not. It was a sovereign decision about who would live and who would die with economic considerations often weighted more heavily than humanitarian concerns. Predictably, if one looks at the geographical spread of the disease, it was most pronounced in the north and progressively less so towards the south. While this quarantine route policy was an advantage to the planters and to the more heavily populated centre and south of the island, it fell heavily upon villagers who lived along the route in the more lightly populated north. The impact of cholera on parts of the north was indicated by the G A of the Northern Province who wrote in 1863 that in part of the province the smallest class of village, containing one to five males increased by 65 per cent between 1829 and 1862.[98]

While the quarantine route was an advantage from the point of view of public health (the body politic), it was a distinct disadvantage from the point of view of sick immigrants who fell by the wayside along the long, lonely stretches of road. In

---

96 Ibid., 122–23c.

97 Although, as pointed out earlier, some carriers can remain asymptomatic for up to two weeks.

98 *SP*, 1867, 18.

order to attempt to balance the demands of the body politic against the bodies of sick labourers, the northern route became a highly bureaucratised space. Permanent and temporary 'cooly hospitals' were established all along the route and in between them were 'cooly sheds.' The hospitals were supervised by Tamil medical practitioners, with a staff of *kanganies* and peons who moved between stations once or twice a week picking up the sick and dead. To assure that proper procedures were followed, medical practitioners were required to send weekly diaries to the Inspector General of Hospitals. These practitioners were in turn supervised by European medical personnel. For this purpose, the route was divided into two parts with the northern half under the Colonial Surgeon of the Northern Province, and the southern half under the Colonial Surgeon of Kandy. Under them in turn were two junior European medical officers who inspected the route monthly.[99] In addition to this, in 1872, all *kanganies* arriving in Ceylon with labour gangs were given a printed form on which the total number in the gang and name of each labourer was printed. Upon arrival at the estates, *kanganies* were required to provide a certificate stating where and in whose hands each missing immigrant was left.[100] But *kanganies* undermined this process as best they could. The AGA of Matale wrote in frustration,

> when any cooly is missing from a *kankanie's* list, the *kankanie* gives whatever excuse his inventive tongue can frame at the moment, without the slightest regard for truth; and, as he cannot be detained until inquiry is made along a hundred miles of road, the lie is only detected when the kankanie is out of reach of the inspector.[101]

As a consequence of such tactics, the certificate scheme failed within two years.[102]

In spite of all of these bureaucratic measures and surveillance of Tamil labourers and *kanganies* by medical practitioners, and of practitioners by other British officials, there was still a fear that there was not enough surveillance. As Dr Kynsey said, 'it can not be denied that the length of the road between the port of debarkation and the coffee districts, offers serious obstacles to supervision.'[103] And yet in spite of these difficulties the surveillance was, in fact, intense as illustrated by the report of the Colonial Surgeon for the Central and North-Central Provinces in 1876. He reported that on 10 May, word was received from two stations along the immigrant road that there were cases of cholera present. The message was relayed to stations further south along the road for inspectors to examine all groups heading south. On the 22 May the Colonial Surgeon began inspecting all stations on the North Road to see if cholera had been reported. Having found no cases, he remained in Anuradhapura until 5 June and then returned to Kandy, re-inspecting every station and 'cooly shed' on the North Road. The Surgeon added that it was remarkable that more cases did not develop, as between 20 May and 6 June, 14,770 immigrants had passed the Dambulla Station on the North Road. On 1 June the Surgeon reported a case in Kandy and immediately ordered the town be closed to immigrant labourers.

---

99 *ARC*, 1876, 123c.
100 *SP*, 1872–1873, viii.
101 *ARC*, 1872, 63.
102 Bastiampillai, *The Administration*, 22.
103 *ARC*, 1876, 124c.

They were then rerouted to their estates on routes that avoided other large towns. In spite of these precautions, he reported that cholera spread in Kandy and on adjoining estates.[104] But the issue of rerouting was complex, as the health of immigrants had to be balanced against that of the local inhabitants. For example, in 1875 the fear that large numbers of immigrants walking through Kandy might spread cholera to residents of the town, led to a proposal that henceforth they would use an alternate route. This plan was rejected by the government 'owing to the fear that the coolies would suffer from exposure if compelled to go by a new route without sufficient shelter.'[105] As we shall see later, when forced to choose between the interests of the plantations in having ready labour and the interests of local villagers, the planters' interests usually came out on top.

In spite of opposition from the Principal Civil Medical Officer, the Colombo route became the second most important after the completion of the railroads to Tuticorin in south India and from Colombo to Kandy. During 1876 60,089 immigrants took this route. Dr. Kynsey viewed the Colombo route with 'dread and alarm', because of the failure of the choke point system of inspection and because Colombo and the West Province were so populous and hence so 'liable to epidemics of cholera and smallpox.'[106] In 1874 cholera spread within Colombo and police were posted at the Peradeniya Bridge to halt immigrants with cholera entering Kandy by train.[107] In 1875 in an attempt to deal with the fear of the spread of cholera from newly arrived immigrants to the residents of Colombo, immigrants that landed from Tuticorin were taken directly from the breakwater in open wagons to a depot outside of the city where they were washed and disinfected before being sent by train to the estates.[108] Dr. Kynsey urged that the Colombo route be closed and the little-used Kalpitiya-Kurunegala route[109] be substituted. [See Map 4.2]. His reasoning was that this route avoided major towns like Puttalam and Kurunegala and had the advantage of running through sparsely occupied county. He further argued that the route could be inexpensively monitored by constructing temporary 'cooly hospitals' and sheds along it at intervals of a day's walk.[110] In the end it was decided that the costs of upgrading this route ruled it out.

In 1877, with both famine conditions and cholera raging in the south of India, the Ceylon Government temporarily stopped immigration at the Indian port of Tuticorin. Much to the annoyance of the Planters' Association, this produced a shortage of labour for the estates, which the Association claimed could have been averted had the government allowed immigrants to come to Ceylon and quarantined them outside Colombo.[111] The Madras Government also petitioned the Ceylon

---

104 Ibid., 128c.

105 *ARC*, 1875, 40.

106 *ARC*, 1876. 124c

107 *ARC*, 1874, 12.

108 *Ceylon Hansard, (CH) Debates of the Ceylon Legislative Council*, 1875–76. (Colombo: Government Printer, 1876), 50.

109 In 1876 only 240 immigrants arrived via the Kalpitiya-Kurunegala route. *ARC* 1876, 122–23c.

110 Ibid., 125c.

111 *PPA*, 1877–78, 64–65.

Government to allow the labourers to come.[112] In that same year the Governor appointed a commission to study a new potential immigrant route via steamer from Tuticorin to Dutch Bay on the central coast, that had the twin advantages of being shorter than the northern route and crossing through sparsely populated space.[113] [See Map 4.2]. The former was important in preserving the health of immigrants while the latter was, of course, a prime consideration to stop the spread of disease. As the viability of the proposed route was assessed, the same system of visibility and supervision adopted for the North Road was applied.[114] However, there was little faith on the part of locals that they would be safe. As the AGA of one of the districts through which the proposed route was to be sited said, 'the people have a great horror of the immigrant cooly, supposing that he carries cholera with him wherever he goes and will have nothing to do with him, unless they are very well paid.'[115] And they had good reason to be concerned given the experience of villagers who had the misfortune to live along the Northern Route. This new route was not adopted for three reasons. First, the proposed costs of all of the medical facilities, personnel, and fresh water counted against it at a time when government revenues were dropping with the decline of coffee. Second, the government officials though whose districts the route was to pass opposed it. And third, the Northern Route remained the cheapest for the labourers and therefore it was preferable to them, in spite of the hardships they faced. Ultimately, short of sealing the Northern Route to traffic, the immigrants themselves were in a position to decide how they came to the estates. It was they who paid for their passage and although the Northern Route was the most dangerous for them it was also the cheapest.

Looking back on the Northern Route after its closure in 1899, Governor West Ridgeway was brutally honest about the sovereign choices that the Government had made in regard to this route.

> It was the Mannar [northern] route that acted to a certain extent as a preventive and saved the estates from frequent visitations of cholera, but when the system on which this route was worked was considered it will be seen that certain districts in the Northern Province were sacrificed ... to secure the safety of the planting districts ... Upon the inhabitants of the districts through which the coolies passed, the results, ... from the constant passage of infected gangs through uninfected villages, were very disastrous. It is not too much to say that the country along the cooly route was depopulated. The abandonment of the sick which was the coolies' safeguard was the villagers' ruin; what the former escaped by constantly moving on the latter retained.[116]

It is difficult to think of clearer examples than this of governmentality, of treating a population as a resource, and of sovereignty – the decision by the state as to who would live and who would die.

---

112 Bastiampillai, *The Administration*, 23.
113 *SP*, 1878, 9.
114 Ibid., 181–96.
115 Ibid., 9.
116 Wesumperuma, *Indian Immigrant*, 47.

Having said this, just as the immigrant labourers resisted governmentality so also did local people. In 1889 the GA of the Western Province reported that Sinhalese villagers were hiding cholera patients from the medical police. He wrote 'the people will not give information of the cases, because they are afraid of being taken to hospital and because they prefer native practice. To prosecute is impossible unless there were an isolated court or prison.'[117] While at times there was spontaneous resistance to British bio-political moves, at other times opposition to anti-cholera measures was highly coordinated. As an example of the latter, the GA of the Eastern Province reported in 1891 that,

> Many of the Vellalas, who form the most influential class here, formed a secret society to work up the poorer and more ignorant classes to oppose sanitation in every way. Finding our patience inexhaustible and the good sense of the lower classes successfully combatted their conspiracy, they finally employed the most ignorant of their own class, Tanakarans with hired ruffians from Jaffna to assault the officials. Notwithstanding this we continued our efforts and towards the close of the epidemic the prejudice of all but the influential conspirators had been broken down.[118]

Such opposition was treated with great care by the British as it could flair up into political unrest as it did in Trincomalee in 1891 when cholera prevention measures caused rioting in the city.[119] Even in the face of death, resistance to biopower was intense.

Sanitation and the Abject Body

According to nineteenth century theories of sanitation there is an intimate connection between smell and disease. Bad smells were seen to signal danger. Just as there could be miasmatic, diseased environments, it was thought that there were bodies that were both bad-smelling and diseased. Bodies were seen as prone to disease because of permanent bad odours. Bodily odours were also thought to pollute the air and spread disease. Miasmatic environmental theory thus had a parallel at the level of the human body. As discussed in Chapter 3, brown bodies could be objects of desire but also objects of disgust and abjection. For example, consider the following by the planter Millie:[120]

> the dark-skinned races of all shades, wash them, clean and scrub them as you will, have naturally an odour proceeding from the skin, which nothing can check or take away; and this odour is naturally repulsive to the sense of smell in a European, and it may—no doubt does—in some measure proceed from this, or may account for the in many cases almost instinctive aversion one feels in sitting down in a railway carriage or any crowded room in close proximity with a black skin. This feeling of aversion does not proceed from the

---

117 *ARC*, 1889, 22.

118 Ibid., 1891, 11.

119 S.A. Meegama, 'Cholera Epidemics and their Control in Ceylon', *Population Studies* 33 (1979): 143–56.

120 Millie, *Thirty*, Ch. 37.

fact that the man's skin is a different colour from yours, but from the exhalations which proceed from it.

This is a classic expression of abjection. It is an especially remarkable statement in the way it naturalises racism. The idea that in some fundamental sense certain human bodies can not be cleaned is appalling or at least extremely odd sounding today, but it was considered reasonable, even scientifically sound by many in the nineteenth century.

Pollution both then and now is defined as the breaching of boundaries between entities, which produces abjection, a visceral disgust. From the perspective of modern sanitation and medicine, pollution entails a breaching of the boundaries of that which rationally must be kept apart. Sanitation concerns protecting the boundaries between what is fit to eat and what is not (the fresh versus the decayed, the nutritious versus the poisonous), and protecting the body from re-ingesting its own or others' waste. I explore these issues with reference to a controversy that flared up in 1869 around a report on the health of estate labourers filed by W.G. Van Dort, the Assistant Colonial Surgeon in charge of the Gampola Hospital that pointed to a 24 per cent death rate among migrant labourers between 1843 and 1867. Acknowledging that rather than actual mortality figures, which were unavailable, this was an estimated percentage derived from immigration returns added to an estimated number of migrants who had settled permanently, he nevertheless concluded:[121]

> Even making a large deduction for incorrect Returns, there is no doubt that the mortality among Malabar coolies is excessively high, and that a large proportion of deaths is due to causes, which a more elaborate system of Medical Police, beginning its supervision from the maritime districts of the Northern Province, and following these Immigrants along the inland route to their several places of destination, together with strict enforcement of sanitary laws in cooly lines in coffee estates, could certainly prevent.

Although, as we have seen, earlier reports on mortality had focused upon the perils of the northern route, Van Dort identified sanitation on the plantations as another principal medical problem. For the planters, this was like waving a red flag before a bull. Although during the 1840s, planters had been blamed for the ill health and deaths of their workers, by the 1850s and 1860s government officials became convinced that the responsibility for the high immigrant death rate lay with the *kanganies* and the coastal advance system. Van Dort's report in turn challenged this view suggesting that unsanitary conditions on the estates were in fact largely responsible.

The doctor's emphasis on sanitation reflects the influence of new developments in public health and sanitation in Britain. In the 1840s Edwin Chadwick was instrumental in making sanitation an important component of social policy in Britain. He urged the creation of a central public health authority to eliminate cesspools and construct sewage removal systems. Chadwick and others, however, were much more reluctant to confront the possibility that poverty itself that was a cause of disease. To admit this would imply that more major structural changes in society were required which

---

121   W.G. Van Dort, 1869–70. 'Gampola Hospital Report 1869', reprinted in *PPA*, 1869–70, 54–55.

would further undermine the still generally laissez-faire approach of government.[122] Chadwick was greatly influenced by Benthamite ideas, and saw sanitary reforms as a vehicle that 'would enable the poor to make full use of their natural capacities, thereby reducing their reliance on public funds.'[123] His views influenced the Royal Commission of Health in Towns in 1845 and the implementation of the British Public Health Act in 1848.[124] Such ideas were reinforced by Florence Nightingale's statistical analyses which confirmed the suspected correlations between disease and sanitation and the subsequent campaign for improved sanitation after the Crimean War in the mid 1850s, which led to a reform of the entire British military hospital system. Further impetus came from the Royal Commission of the Sanitary State of the Army in India in 1863 which found that living conditions in the Indian army were the principal cause of disease, thus shifting responsibility for ill-health away from the tropical climate towards human negligence.[125] Although the role played by bacteria in disease was not yet understood, Europeans were granted a measure of protection from many diseases by their increased concern with personal hygiene and their belief in the importance of clean drinking water.[126]

The idea of 'dirt' became increasingly prominent in colonial discourse during the nineteenth century, as the British marked their own hygienic practices off from what they saw as the dirt and disorder of colonial environments.[127] Dirt was not just pollution in the environment, it was thought to emanate from unhygienic or immoral bodies and pollute the surrounding environment.[128] Such conceptions consistently blurred anatomical and geographical space.[129] Sanitarians compared poor and unsanitary areas of British industrial cities to 'darkest Africa' or 'teeming Asia' thus blurring colonial and metropolitan spaces.[130] By 'nativizing' the poor, they maintained a critical distance from abject 'Others' at home. The term 'civilization' was increasingly linked to advances in hygiene and the spread of sanitation.[131]

---

122 J.M. Eyler, 'The Sick Poor and the State: Arthur Newsholme on Poverty, Disease, and Responsibility', in *Framing Disease*, edited by C.E. Rosenberg and J. Golden (New Brunswick: Rutgers University Press, 1992) 279–81.

123 M. Harrison, *Disease and the Modern World* (Cambridge: Polity, 2004), 111.

124 R. Porter, *The Greatest Benefit to Mankind: A Medical History of Humanity from Antiquity to the Present* (London: Harper-Collins, 1997), 411.

125 Collingham, *Imperial Bodies*, 165.

126 Arnold, *Colonizing*, 166. The notion of clean water, meant free from impurities which were understood in terms of chemical composition rather than bacteria. Elaborate filtration was increasingly used which had the unintended result of removing certain bacteria which caused disease. (See Curtin, *Death By Migration*, 50–53).

127 On sanitary regulation in colonial Madras, see M.S. Kumar, 'The Evolution of the Spatial Ordering of Colonial Madras', in *Post-Colonial Geographies*, edited by A. Blunt and C. McEwan (New York: Continuum, 2002), 85–98.

128 A. Bashford, *Purity and Pollution: Gender, Embodiment and Victorian Medicine* (London: Macmillan, 1998), 18.

129 D. Armstrong, 'Public Health Spaces and the Fabrication of Identity', *Sociology* 27 (1993): 393–410.

130 Worboys, 'Germs', 184.

131 Bewell, *Romanticism*, 42, 251.

One reason why contagionist theories of disease were opposed by sanitarians in Britain and South Asia is that they saw disease as a medico-moral complex while contagionist theories were in the words of Rosenberg, 'morally random.'[132]

Dr. Van Dort's report placed the blame for the high death rates on conditions on the estates. However, it was the labourers themselves whom he saw as largely responsible for these conditions. 'Habits of life', he said, 'may be considered as bearing ... [an] ... important share in inducing the diseases to which the Malabar cooly is most liable. The immoderate indulgence of his gross and sensual appetites, his want of cleanly habits, his improvident and careless disposition, his apathy and indolence, render him an easy prey to disease.'[133] Van Dort singled out the labourers' eating habits as a mark of particular abjection.

> The indulgence of their animal appetites is their sole motive for exertion. They eat to repletion, and are far from discriminating in the choice of their food. There is indeed nothing short of poison, with which a low caste Malabar will not try to appease his hunger. Putrid meat, entrails, reeking hides, roots, leaves, nothing comes amiss to him. He will exhume the buried carcase of a sheep to feast on its decomposed remains.[134]

For Van Dort, the labourers behaved in ways that rendered them susceptible to disease. They literally polluted themselves. On the other hand, it is clear that Van Dort thought that the labourers' pre-modern culture diminished their responsibility for what tradition had wrought. Van Dort's description of them, their racialized bodies as well as their practices, stands as a catalogue of all the qualities that make them, an abject Other. This view was reaffirmed by the AGA of Nuwara Eliya who reported planter complaints that immigrants' 'mental faculties are but little superior to those of irrational animals ...'.[135] While it is clear that these charges of animality are suffused with implicit claims of racial difference, the same charges were made against paupers in Britain, revealing how class and race substitute for and slip into one another. For example, Chadwick in his *Report on the Sanitary Condition of the Labouring Population of Great Britain in 1842* presents the following testimony:

> the bone-pickers are the dirtiest of all the inmates in our workhouse; I have seen them take a bone from a dung-heap, and gnaw it while reeking hot with the fermentation of decay. ... they were thoroughly debased. Often hardly human in appearance, they had neither human tastes or sympathies, not even human sensations, for they revelled in the filth which is grateful to dogs, and other lower animals, and which to our apprehensions is redolent only of nausea and abomination.[136]

132 C.E. Rosenberg, *Healing and History* (New York: Science History Publications, 1979), 17.

133 Van Dort, 'Gampola Hospital', 57–58.

134 Ibid., 56.

135 *ARC*, 1872, 71.

136 E. Chadwick, *Report on the Sanitary Condition of the Labouring Population of Gt Britain* (Edinburgh: Edinburgh University Press, 1965 [1842]), 164–65, quoted in Eyler, 'The sick poor', 286.

The implication of Van Dort's report was that the Tamil labourers could not be trusted to attend to their own interests. They needed constant surveillance and discipline so they could be protected from themselves. They were thus seen as prime candidates for programmes of governmentality to be carried out on the estates under orders of the state. The planters resented Van Dort's accusations of neglecting their responsibilities. They strongly resisted being drawn into state bio-political objectives and furthermore denied that a major health problem existed.

The Emigration Board, which had responsibility for Indian labourers, received all government reports pertaining to them. Although greatly troubled by the report, the Board was unwilling to accept the validity of the 24 per cent mortality rate, estimating it at closer to seven per cent The Board did, however, appear to accept Van Dort's view that the Tamils were irresponsible by nature. It wrote that their 'unhealthy and immoral mode of life, extraordinary and repulsive as it is, corresponds with the accounts formerly received from the West Indies, respecting coolies from Madras, which led at last to the discontinuance of emigration from that Port to the West Indies.'[137] The Emigration Commissioner[138] concluded that rather than allowing

> the people to destroy their health by feeding on garbage, the best plan would be to allow the employer to ration the labourers, making a proportionate reduction from his money wages. It would be necessary in that case to give the Government the power of prescribing the amount of ration, and of fixing the deduction from wages to be allowed on account of it. I presume that the Government would be able through the Agents of the several Provinces to watch sufficiently over the Coolies, to see that they were not defrauded, either in the rations issued to them, or the amount deducted on account of them.

The thought of government interference in the running of the plantations was anathema to the planters on financial as well as ideological grounds.[139] Although keen to encourage assistance when it profited them, they played the laissez-faire card whenever they felt their profits were threatened. Nevertheless, planters were fearful of the power of the Commissioners and concerned that they 'would be looked upon as sort of Legrees.'[140] Hence the Planters' Association set out to demolish Van Dort's arguments. The Secretary of the Planters' Association and several senior planters responded by undermining Van Dort's most damaging charge; the mortality figure of 24.6 per cent. First, they argued that, as there were no reliable statistics collected on immigration or emigration or mortality, no credence could be put in Van Dort's figures.[141] Second, they pointed out that even if one accepted the statistics, his use of

---

137 Murdoch T.W.C. to Sir F. Rogers reprinted in *PPA*, 1869–70, 64–68.

138 Ibid., 68.

139 Public health intervention to many early Victorians was thought to be an authoritarian intrusion on human freedom and the workings of 'economic laws', Porter, *Health*, 112.

140 *PPA*, 1870–71, 2. Planters had anxieties about being associated with the West Indian slave-owning planters of the early nineteenth century. The slave planters were held, especially by Evangelicals to be a very brutal and degraded type of Englishmen. On English perceptions of slave owning planters see Hall, *White*, 208.

141 The first Ceylon Census was not until 1871. Before that time there was great uncertainty about the extent of the state's resources and liabilities. In an age when statistics were seen as very important for understanding and managing the world, this was a serious

them was seriously flawed. And third they presented their own surveys of the estates revealing (so they claimed) average rates of mortality to be around three per cent for 1868 and 1869. Finally they attempted to discredit his characterisation of the labourers. While Van Dort had suggested that the labourers were forced of necessity to eat garbage and that the planters had failed to supervise their eating habits, the planters responded by charging Van Dort with such ignorance of the estates that he mistook the exception for the norm. 'It is true', the Secretary said,

> that there are amongst them some very coarse feeders, but those form an insignificant and entirely exceptional few. And whilst it is also a fact, that, in certain very rare individual cases, they have been known to eat the disgusting substances mentioned by Dr. Van Dort. But even in these cases, it is rarely a simple hunger that impels them, but the cravings of a disordered system.[142]

---

shortcoming. Although the state made numerous attempts at gathering statistics on bio-political issues such as births and deaths, rates of illness, and migration, the figures collected were viewed with little confidence. An important part of the problem of obtaining accurate statistics was the natives' resistance to the collection of information on them. The AGA of Matale in 1867 wrote of the 'utterly worthless character of statistical returns in Ceylon ...' which was due to 'the suspicions with which Natives view all statistical inquires by the Government ...' (*ARC*, 1867, 30) The AGA of Nuwara Eliya put down this resistance to the 'natural aversion of the Sinhalese to innovation of any kind.'(Ibid., 51) However, it was not only the non-Europeans who failed to report information. The GA of the Central Province, wrote in his annual report of 1867 that 'a large proportion of the deaths amongst coolies employed on estates or travelling to or from the continent of India are not reported at all to the registrars. Circulars have been sent to Estate managers ... asking them to report births and deaths which might occur amongst their coolies, but very few comply with the request, nor does there exist any effectual means of securing such returns.'(ibid., 23) There was in fact a means of forcing compliance. It was Ordinance 17 of 1862 which made it a criminal offence not to report the mortality of natives of India employed in Ceylon. The AGA of Matale in 1871stated that this avenue was not pursued 'through a disinclination on the part of Government officials to prosecute criminally gentlemen of respectability ...' (*ARC*, 1871, 55). In fact, from the start, planters refused to cooperate with what they considered to be an intrusion into their business. Governor MacCarthy felt that the planters were powerful enough as a group that he could not prosecute them as a whole if they chose not to cooperate. Consequently the law remained on the books but unenforced (Vanden Driesen, *Indian Plantation Labour*, 95). The AGA went on to say that the returns on the numbers of labourers working on individual estates were ridiculous. He cited one estate that listed employing only two male labourers and another that listed 2,088, which was subsequently upon being queried corrected to 37 (*ARC*, 1871, 55). The following year, the Government decided to begin prosecuting planters who failed to post returns. After numerous prosecutions, the Government was more confident that they were beginning to get a sense of the extent and condition of the labourers. However, the AGA for Matale wrote of planter resistance to this collection of statistics by filing returns 'which they knew to be incorrect, and afterwards retailing the "joke" for the edification of their fellow-countrymen!' (*ARC*, 1872, 63). The planters could, however, produce statistics on labourer mortality when it suited them. For example, they produced figures for 1868–69 in order to rebut medical charges of negligence on their part (*PPA*, 1869–70, 81).

142 Byrde, H. to The Colonial Secretary, reprinted in *PPA*, 1869–70, 78–83.

The problem, another planter suggested, was that Van Dort saw only the labourers who had failed physically and morally. 'The doctor's experience', he argued, 'is mainly drawn from contact with the filthy, the vicious, the reckless, the drunken, and the lazy, (laziness of course often entailing destitution and suffering).'[143]

The planters' denials were crucial to the planter's defence, for if the planters were proven to have failed in their responsibilities, they would have to be monitored by the government and bear the financial cost of that surveillance. Most importantly they wished to avoid a ban on immigration which would cause the collapse of the whole coffee industry. More realistically, however, they feared losing control over food, sanitation and health provision on the estates.

In his response to Van Dort, the Secretary of the Planters's Association concluded that labourers could afford to purchase their own food and that rations need not be provided as part of their pay.[144] Some of the supporting letters from other planters shed light on the planters' attitudes. One made the argument that providing rations 'would probably cause the daily rate [of pay] to be higher than at present, which Planters can ill afford, and would also lead to much annoyance as coolies would feign illness, and endeavour to get rations without working.'[145] Others made the case against providing food as part of pay on the grounds that if people 'are so thoroughly cared for ... they lose eventually all habits and power of caring for themselves, and when they get a little cash in hand, prove themselves too often utterly incapable of all self-restraint.'[146] Still others played the powerful anti-slavery card arguing that '[i]f rations should form a part of wages, work must be had for them, and a state of things would be brought about approaching to the horrors of slavery.'[147]

While most of the debate surrounded questions of food and health, Van Dort's report touched on broader issues of sanitation on estates. Not only did the planters want to avoid having the government usurp their disciplinary powers within the boundaries of the estates, they appealed to the government for more power to discipline labourers. The Association suggested that 'the government might make punishable the eating of offal and putrid meat, and there should also be a law to enable planters to have coolies punished if not using "latrines" when provided for them.'[148] The fact that many labourers were unaccustomed to latrines was used to justify not building them.[149]

Time and again, the planters portrayed themselves as suitable guardians of the health of the labourers within the estates. 'As a general rule', one wrote, 'coolies are successfully treated for fever and dysentery by the superintendents of estates, and if coolies die it is generally from want of attention, on the part of their fellow labourers, or from bathing in cold water or eating pork, or doing something which

---

143   J. Smith to H. Byrde reprinted in *PPA*, 1869–70, 99.

144   H. Byrde to The Colonial Secretary, reprinted in ibid., 83. Labourers purchased rice from the estates on which they worked.

145   W. Sabonadiere to H. Byrde, reprinted in ibid., 98.

146   W.A. Swan to H. Byrde, reprinted in ibid., 103.

147   P. Moir to H. Byrde, reprinted in ibid., 96.

148   Sabonadiere, *The Coffee Planter*, 98.

149   Wesumperuma, *Indian Immigrant*, 234.

brings on a relapse and so carries them off.'[150] In fact the major killers on plantations were colitis, tuberculosis ulceration, malaria and food poisoning often from eating harmful tubers and roots. Labourers also died of scrub typhus, contracted from mites during the opening of new land. Hookworm was also a major problem associated with the unsanitary conditions of plantations, but it tended to debilitate rather than kill.[151] Many of these diseases were untreatable by trained doctors, let alone untrained planters.

The Van Dort report in many respects marked the end of the defacto sovereignty of the estates. It became harder for planters to pursue their policy of trying to merely 'make survive' in an environment that was not conducive to survival. In spite of their spirited defence against the report, the planters never really regained the autonomy over labourer health that they had. However, as we shall see, the planters remained a powerful force on the island and used every resource at their disposal to fight a rear-guard action against what they saw as the usurpation of their rights as economic subjects to control their labourers. It is important not to see Van Dort simply as an individual. As the director of a government hospital, he was part of an institutional system of biopower that was aggressively colonising both metropolitan and colonial society. Colonial hospitals were forced to confront the material fact that immigrant labourers had a mortality rate that was much higher than that of other groups on the island. As part of the government bureaucracy, hospitals were required to keep records on admissions and deaths and justify them to the government annually. They had an interest in showing that the high mortality among Tamil labourers was caused not by inferior care in the hospitals, but by their abject condition before they reached the hospitals.

GAs and AGAs in the planting districts also had to justify the mortality of the population under their control. In his report for 1871, the AGA of Matale blamed the high death rate (1 in 3.78) amongst labourers in the Matale Hospital on the late stage of illness in which they entered hospital. He wrote that there is

> a mistaken idea in the minds of some superintendents that the cooly should be allowed his own choice, whether he should be sent to the hospital or allowed to remain and die on the estate ... A surgeon is not to be blamed for unskilfulness when none but dying patients are entrusted to his charge.[152]

He stated that labourers' health problems should not be left to an 'unprofessional superintendent', and proposed a system of district hospitals.[153] The bureaucratic structure of the hospitals and government combined to put pressure on the estates to assume their bio-political responsibilities. The Planters' Association resisted this governmentality in the name of liberalism. Their real concern, however, was the cost to them of a government-imposed medical programme. Planters argued further that labourers had a right not to use western medicine. In making their case, planters

150 Sabonadiere, *The Coffee Planter*, 97.
151 Vanden Driesen, *Indian Plantation Labour*, 16, 57.
152 *ARC*, 1871, 58.
153 Ibid., 58.

spoke of the failure of labourers to go to hospitals because of caste prejudice.[154] But the planters did not speak as one. Some, such as James Taylor, argued that it would be good to have doctors, especially European doctors available in the planting districts, as planters could avail themselves of this resource.[155] This view was echoed by another planter who argued that 'our present medical man, being a native of Ceylon, ...[is]... fully competent to treat all ordinary cooly cases, but to meet the requirements of the European portion of the community as well, a higher class man might be introduced.'[156] Others such as the planter Shand argued that what was needed was:

> a class of educated men who, possessing the affinity of color and language, can frequently visit our lines, can point out the advantages of cleanliness, the errors of filth, of clay eating, of carrion eating, of procuring abortions, of cohabitation among near relations, and of many other existing habits which tend to swell our mortality and which an European doctor, who would simply deal with the effect rather than striking at the cause, would be powerless to check.[157]

Of course, the planters were well aware that non-European doctors would cost them less. It was estimated that whereas the cost of dispensaries and attendants would be the same for a European, a Burgher or a native doctor, the cost of salaries and housing would vary considerably. A European would draw a salary of 4,000 rupees, and get a house worth 2000 rupees, a Burgher a salary of 1500 rupees and a house worth 600 rupees, and a native a salary of 1000 rupees and a house worth 600 rupees.[158]

Two years after the Van Dort report in 1872, Governor Gregory in his opening address to the Legislative Council stated that '[m]uch and deserved interest is felt in England on the subject [of immigrant labour]. Public feeling has on more than one occasion been aroused by statements of ill-treatment to which immigrant labourers have been subjected in British possessions.' He went on to say that upon personal inspection of the immigrant routes and the conditions on the plantations that he was satisfied that labourers were being well cared for. However, he added that he had been' pressed by the Imperial Government' in light of the success of the Medical Charities Act in Britain to propose a medical ordinance for the protection of labourers on estates.[159]

Consequently, the government in consultation with the Planters' Association put forward the following ordinance. District medical committees were to be elected by the estate proprietors to oversee the medical needs of the district. If the proprietors failed to appoint a committee, the government would make all of the medical arrangements itself. The planters would be assessed at the rate of one rupee per acre. Medical officers were to be appointed by the governor, but salaries were to be fixed by the district committees. Superintendents were to provide access to the

---

154 *PPA*, 1871–72, 139.
155 Ibid., 148.
156 Ibid., 291.
157 *PPA*, 1872–73, 91.
158 *PPA*, 1871–72, 176–77.
159 *SP*, 1872–1873, vii–viii.

lines for inspectors, provide rooms for sick labourers, remove the sick to the central hospital at the proprietor of the estate's expense, upon the recommendation of the medical officer, and keep registers of births and deaths.[160] The proposed ordinance was hotly debated in the Legislative Assembly. The planters realised that they could not stop the ordinance, but set out to weaken it. The principal strategy was to campaign for as much scope as possible to enforce the ordinance themselves. As the representative for general European Interests said, 'more power must be given to them [the planters] by the Government. An Ordinance of that kind should be framed with very free and liberal provisions, retaining the power to the Government to step in if there was any irregularity.'[161] He argued that the planters themselves should be the ones to choose the doctors rather than the Civil Medical Officer.[162] The Tamil representative on the Council, countered that although 'no man depreciates unnecessary Government interference more than I do', the plan to let planters appoint medical committees would give planters too much power.[163] He also tackled head on the issue of paternalism versus liberalism and the labourers' aversion to western medicine. Should a labourer be compelled to go to hospital if he didn't wish to go? Yes, 'if anything justified the slightest restriction on any man's liberty, it was the desire to prevent him from enduring the sufferings of disease.'

The Auditor General argued that, while he was content to have the planters appoint the medical committees, the government must appoint the doctors. This he said was based on his experience in Mauritius where planters abused the system by appointing doctors who did not enforce the medical ordinance and consequently the whole medical ordinance broke down.[164] The Colonial Secretary pointed out that the proposal did not constitute an unusual level of government interference as the Government of India had instituted a similar ordinance for their tea plantations, and that an even more stringent ordinance had been enacted in Ireland.[165] He was adamant that the government would appoint medical officers 'that should be responsible to Government for the proper performance of their duties, and should be liable to summary dismissal, if they neglected their duty either to the Planters or to the coolies.'[166] He continued,

[i]t was difficult to see for example what check such a committee would exercise in a case in which the Doctor by connivance of a Superintendent of an estate indifferent as to the treatment of coolies, should shirk his duty. It was absolutely essential that the medical officer should be a servant of the Government, subject to its inspection and control.[167]

Nevertheless, the government subsequently caved in to planter pressure and allowed the planters to appoint the doctors. The Medical Ordinance was brought to a vote

---

160 *CH*, 1872–73, 2.
161 Ibid., 44.
162 Ibid., 44.
163 Ibid., 45.
164 Ibid., 46.
165 Ibid., 47.
166 Ibid., 48.
167 Ibid., 48.

and passed eleven to three, the European unofficial members voting against the ordinance, even though they had managed to strip it of most of its power.[168]

The state had intervened in this very important bio-political area, on three grounds. First, because it was felt that the labourers were ignorant of sanitation and that their antipathy towards western medicine was such that they could not be trusted to seek 'proper' medical help. Although there can be little doubt that there was antipathy to European medicine, this antipathy provided an excuse for planters not to 'waste money' on worker health care. Second, in the climate of increasing bureaucratisation and scientifically-based professionalisation, laymen were no longer seen as credible guardians of health, even for those deemed to be abject. In fact, one planter expressed his support of scientific bureaucracy in the service of the greater good:

> It is absurd and inhuman [he wrote] to expect superintendents of estates to form a correct diagnosis of one-tenth the diseases current among Tamil coolies. Guides published by doctors are a mere fallacy; not one in a hundred corresponds with the symptoms as given. That the lives of nearly 300 coolies should be left to the medical charge of a person so thoroughly ignorant of the effects of medicine as myself is disgraceful. Any person who may argue to the contrary I shall be most happy to treat the next time they are suffering from serious illness. What then is the objection [to the proposed ordinance]? Surely not thousands of human lives in the balance with a few pounds or rupees or cents.[169]

Although officials were cautious not to appear too openly critical of planters, it was felt that many planters did not have sufficient control over labour on their estates to adequately be responsible for their health. Planters contributed to this impression when they argued that they could not always protect labourers from doing themselves harm or from being harmed or neglected by *kanganies*. The planter James Taylor wrote that 'for myself I do not hear of more than half the deaths on this place, directly and at the time ... It would seem as if the labourers often objected to tell of a death.' He went on to add that he offered medicine to labourers when he found them ill but that 'still half the cases of sickness in the lines are not reported to me. Were I to interfere more, they would just take more care to hide cases of illness ...'[170] One planter who was willing to speak up on the subject agreed that sick labourers lives were largely in the hands of *kanganies*.[171]

> If the cooly is in debt to the *kankani* or a relation, he stands far better chances of recovery. The *kankani* calls the attention of the manager to it in either case: "That man owes me a lot of money; if he dies I shall be a great loser; do master, please try and save him." If none of the above causes interest the *kankani* in him, it often is "he is unwell," lazy, or something very trivial, till the Superintendent is surprised by the request of a couple of *mamoties* [hoes] to bury him.

In 1875 after the ordinance had been in effect for two years, the Colonial Secretary asked the Medical Inspector of the Coffee Districts for a report on the extent of its

---

168 Ibid., 49.
169 Ibid., 44.
170 *PPA*, 1871–72, 262–63
171 *SP*, 1872–73, 45.

adoption.[172] The Medical Inspector reported that in some districts the ordinance had still not been even partially implemented.[173] He wrote:

> a feeling, I regret to report, seems to have more or less prevailed that, beyond the appointment of a Medical Officer and the building of a District Hospital, Government would not insist upon the other clauses of the Ordinance … and I anticipate much opposition will be offered in certain districts to … the establishment of sick lines on every estate.[174]

The planters wanted to convert a few rooms in the lines into sick rooms, a cost saving measure which the inspector considered, 'would be quite incompatible with the proper treatment of the sick.' A much more serious problem, he pointed out, was the fact that the planters were using the government doctors nearly exclusively for themselves. This was the consequence, he argued, of allowing the planters to appoint the doctors, for 'if Government looks to them for carrying out the most important provisions of the ordinance, then, in my humble opinion, they ought to be solely responsible to Government.'[175] He concluded, damningly for the planters, that:

> there are, more particularly in the higher regions, very few endemic causes of disease indeed beyond those arising from bad accommodation, contaminated water, the insanitary condition of the lines generally, and perhaps the penurious habits of coolies themselves as regards the purchase of nitrogenous food.[176]

However, the power of the planters was such that they managed to have the Medical Inspector's recommendation that separate sick lines be constructed overturned.[177] Furthermore, the government did nothing to stop the planters from turning the system of health care for labourers into a private care system for themselves. Nevertheless, the Planters' Association expressed general dissatisfaction with the ordinance.[178] They apparently felt justified in resisting all forms of governmentality that were not in their own economic interests. As Vanden Driesen points out, the planters were a very powerful force on the island. Not only did they have powerful political organisations, the Planters' Association and the Chamber of Commerce, they were also represented on the Legislative Council, had great influence with the local press and powerful political friends in Britain.[179]

In spite of having thoroughly undermined the ordinance, the planters driven by their declining fortunes due to the ravages of leaf disease, were determined to weaken the ordinance still further. In 1878 the Planters' Association asked the

---

172 *SP*, 1875–76, 167.

173 *SP*, 1875–76, 169–78.

174 In 1876 there were 22 district hospitals in the planting districts to treat plantation labourers.

175 *SP*, 1875–76, 170.

176 Ibid., 169.

177 *PPA*, 1875–76, xvi.

178 Ibid., 247–62.

179 Vanden Driesen, *Indian Plantation Labour*, 68.

government to appoint a Commission of Inquiry into the Medical Ordinance.[180] They argued that the ordinance was too expensive and must be modified. Some planters wanted to return to planter autonomy. As one said before the meeting of the Planters' Association, the ordinance could not work effectively because European doctors were too expensive an option to treat plantation labourers, and native doctors (trained in western medicine) were 'terribly deficient.' The speaker concluded that thus the planters were:

> compelled to fall back upon what was sneeringly called the patriarchal system – the treatment of the coolies for simple ailments by the planters themselves – which he believed the experience of the Medical Ordinance had taught most of them was the only scheme that would satisfactorily work ... the Government was perfectly right in stepping in and making provision for cases of child-birth and surgical cases, which superintendents could not deal with; but beyond this, Government had no right to go.[181]

Others took the contrary position, wishing to have the government take over the medical care completely, but at a charge of less than the current one rupee per acre. The annual report of the Medical Inspector for Coffee Districts for 1878 was also very critical of the operation of the ordinance. Much of his criticism was directed at the planters who in his opinion subverted the ordinance to their own purposes. He wrote:

> no small amount of the unpopularity of the Ordinance and the want of success in its working have arisen from differences between superintendents and District Surgeons as regards private practice, and the coolies, as even the numerically defective returns of deaths themselves indicate, have been grossly neglected in their illnesses, and the legitimate object of the ordinance – viz., the medical care and treatment of the Indian immigrant labourer – has been diverted into a totally illegitimate channel – the medical care and treatment of the European residents ... More than one half of the total assessment is absorbed by the salaries of the European medical officers, leaving alone the cost and upkeep of the excellent bungalows provided for them, contrasting strangely with the penthouses set apart for the sick coolies.[182]

In 1878 the Governor duly appointed a commission to draw up a new ordinance, but this retained the basic structure of the earlier ordinance and was opposed in Council by the planting member and European member for commercial interests. The G A of the Western Province stated in Council that the old ordinance had failed because the planters did not support it. He pointed out that while it had been intended to supplement the existing voluntary medical aid previously provided by the planters, they often ceased giving any medicine on the grounds that it was no longer their obligation to do so. He went on to state that, 'the object of the commissioners [in framing a new ordinance] had been to bring back a system which would allow as little interference on the part of Government so as to make the scheme as cheap as

---

180 *PPA*, 1878–79, 20.

181 Ibid., 22–23.

182 *SP*, 1879, 541. Behal, 'Tea and Money', p. 158, documents how tea planters in Assam behaved in very similar ways.

possible and secure the interest of the planters in the coolies.' He suggested that the only way to 'secure the success of the Ordinance ... [was to] ... be certain of getting the hearty cooperation of the European residents on the estates.[183] And the Governor agreed. As we can see from the way the government yielded time and again to planter demands, all sides knew that power was fragile and highly negotiable and that the bottom line tended to be the costs to the plantation economy.

In 1880 the Principal Civil Medical Officer submitted to the Council a framework for a new ordinance. It was to be run entirely by the government with an assessment on each estate to help defray a portion of the cost. The officer felt that this was the only workable plan financially, as any other system would be 'beyond the present resources of the planters, who have had a succession of unfavourable seasons with deficient crops.'[184] His plan entailed superintendents treating mild cases of disease on estates and either summoning a doctor or sending sick labourers to hospital. In addition to the general assessment, planters would be charged for doctor's visits to the lines and for the first month that a labourer was in hospital. If a sick labourer was thrown off his estate by a *kangany*, the estate at which he was last employed would be liable for the costs. It was estimated that the cost of this plan would be 75 cents per acre plus the cost of bringing in a doctor and hospital stays. The Principal Civil Medical Officer argued that such a system of sliding costs would encourage planters to hire healthier labourers and look after them better. Medical officers would henceforth be chosen by the government and would only treat planters in their private practice. Furthermore, the government would contribute to the scheme as government labourers and paupers would also avail themselves of the hospitals.[185]

A subcommittee of the Planters' Association also presented a report arguing that the plan was extravagant. They argued, 'in no country can the medical relief afforded to the working classes be of the best.'[186] When the ordinance came up for a first reading in Council in late 1880, the Lieutenant Governor outlined its provisions. The only significant change from the Principal Civil Medical Officer's plan was that rather than a charge per acre, the costs of the system would be borne by a 20 per cent per hundredweight export duty on coffee and other new plantation crops such as tea and cinchona.[187] At the second reading the member for the planting community opposed it, arguing that the government should extend health care to all labourers on the island and that to suggest that only the plantation labourers needed such care was a slur on the planting community. As such, he argued that the ordinance was a special tax on coffee planters who were already greatly suffering financially due to the ravages of coffee disease. Instead of an export tax on coffee, he proposed an import tax on rice to pay for a programme that would not only benefit labourers on estates, but those in government employ and Sinhalese peasants living near hospitals as well. The Auditor General replied that there could be no question of raising such a tax, falling as it would, disproportionately upon the poor. The Sinhalese member

---

183 *CH*, 1879, 132.
184 *SP*, 1880, 81.
185 Ibid., 81–86.
186 Ibid., 305.
187 *CH*, 1880, 52–54.

then spoke against the proposals of the planting community saying that the latter was portraying the export duty as a new tax when in fact it was simply replacing the acreage assessment. He stated that the planters were attempting to 'shirk the payment of the tax which they had hitherto been paying, and to throw the burden of it ... from their own shoulders upon those of the general community.'[188] He also took issue with the claim that the ordinance was for the benefit of Sinhalese peasants as well. He said, '[i]f the Sinhalese labourers who were scattered over the rural parts of the country, were to have the same amount of aid as the Tamil labourers, instead of R. 150,000 being sufficient to meet the purpose, the amount required would, he thought, be nearer ten times as much.'[189] As for the proposed import tax on rice, he stated that 'the people upon whom that tax would fall hardest were the poorest people.'[190]

The Tamil member also refused to support an import duty on rice. The Lieutenant Governor called for this amendment to be defeated, which it was 12 to 3. It was sent to committee but the Planting member refused to sign saying that 'if this bill becomes law it will sever that unanimity and good feeling which has so long and so beneficially existed between the Government and the planting community.'[191] He then handed in a protest to be sent to the Secretary of State calling the export tax 'unconstitutional.'[192] Nevertheless, Ordinance 17 of 1880 was passed into law.

Ultimately, the new ordinance was rejected by the Home Government, on the grounds that London was in principle not in favour of export duties. A capitation tax was suggested instead. Consequently, the following year the Legislative Council proposed amending the ordinance yet again. Emboldened by the rejection of the last ordinance, the planting member arose again to oppose it arguing against 'special taxation in any form'.[193] He stated that the failure of the Ordinance of 1872 had demonstrated that any such ordinance was unworkable and unnecessary and that the government was requiring that the planter provide more health care for labourers than the government provided to Sinhalese villagers. The planters' representative spoke repeatedly of 'official interference.'

> We have come to the Government again and again and told them that the burden laid upon us [the planters] was greater than we could bear. It has been proved to Government that official interference has done much to frustrate the success of the scheme and what do you do? You come to us today to increase the weight of the burden.[194]

He then referred to the capitation scheme as 'the most obnoxious and the most unpopular [plan] which has yet been brought forwards ...'.[195] The Colonial Secretary replied that it was the Home Government that suggested the capitation tax and that

---

188 Ibid., 96.
189 Ibid., 97.
190 Ibid., 97.
191 Ibid., 126.
192 Ibid., 126
193 *CH*, 1881, 52.
194 Ibid., 54.
195 Ibid., 52.

the planters could not shirk their duty to look after the health of their employees. He argued that these are the 'conditions [under which] Her Majesty's Government will permit Tamil immigrant labourers to be employed here.'[196] When the bill came up for a vote, Ordinance 18 of 1881was passed despite the opposition. But the planters refused to give up. Later that year they unsuccessfully petitioned the Queen to overturn the provisions of the Ordinance requiring them to bear the cost.[197]

Opposition to the new ordinance was not confined to the planters. The GA for the Central Province in his report for 1881 wrote:

> no one knows better than I do the vexation and annoyance that the Ordinance of 1872 has caused, not alone the planters, from whom the tax has to be collected, but to every person employed in its collection. But whatever the experience of the past may be, they will, I am sure, be found to be as nothing, when compared with collecting the tax under the present Ordinance.

He claimed that the labourers' health was better cared for before the Ordinance of 1872 and expressed his sympathies for the planters:

> these are hard times, and no one knows better than the tax collector where taxes press heavily. I know that with many it has not been that 'We won't pay the tax', but that 'We can't.' I trust that the day is not far off when the European planter, whose wealth and energy has done so much in times past to develop the resources of the country, will be relieved of this oppressive burden.[198]

The following year the government pronounced the 1881 Ordinance unworkable for the reasons outlined by the GA of the Central Province, proposing an amendment that drastically reduced the costs to planters. This included the replacement of expensive European doctors with cheaper native practitioners 'conversant with the language, the manners, the feelings, and even the prejudices of the coolies.' The government agreed to pay the cost of doctors and of treating labourers at civil medical hospitals.[199] The Colonial Secretary held that the government had not abandoned its views about planters' legal obligations, but thought that the government should consider 'the present depressed state of the planting industry.'[200] He, therefore, decided to abandon his objection to export duties as they were the only practical way of securing a contribution from the collapsing coffee estates.

In 1882 a new ordinance was passed over the objection of the planters' representative. In the current state of financial collapse, the planters considered the government's concessions to be insufficient.[201] They appealed it to the government once again, but in late 1884 became resigned to the idea that the government would not remove the duty on export coffee.[202] By 1886, having abandoned hope of

196 Ibid., 56.
197 *PPA*, 1881–82, 164–67.
198 *ARC*, 1881, 58.
199 *SP*, 1882–83, x.
200 *CH*, 1882–83, 84–85.
201 *PPA*, 1882–83, xx.
202 *PPA*, 1884–85, xvii.

overturning the ordinance, the planters unsuccessfully petitioned the government to employ more European doctors who would focus more of their attention on the planters, leaving the care of labourers and inspections of the sanitary conditions to less highly trained, native practitioners. As one planter put it at an Association meeting, 'we agreed unanimously here that it is a perfect farce to send a highly talented and educated medical man to do what I may call scavenging our lines.'[203] The language of abjection and the metaphor of garbage are revealing. Another stated that 'the Government have taken away from the Planting community the power of helping themselves.'[204] The language here is that of liberalism in the face of paternalistic governmentality. By 1892 the production of coffee was negligible and the coffee period had been effectively over for several years.

As it turned out the 1880s was not a good decade to test the various medical ordinances and their effects because these were years of ecological and financial collapse in the highlands, which took their toll on the health of the labourers as well as the peasantry. In 1892, Governor Havelock appointed a commission of inquiry in response to continuing high rates of death among immigrant labourers. The commission found that between 1883 and 1892, 20.87 per cent of the immigrants who were admitted to district hospitals died as opposed to 10 per cent of admissions to non-immigrant hospitals.[205] It would appear that the various attempts at expanding governmentality had made relatively little difference throughout the coffee period. And yet, the reasons for this failure are relatively clear. Although government framed some reasonable legislation, the planters persistently undermined it. Any legislation that was eventually passed was then effectively gutted by the planters, who used their power and organizational skills to accomplish in the hills what they had been unable to achieve in the Legislative Council. For example, just as planters threatened labourers who took them to court over unpaid wages with physical violence, so they threatened government officials who sought to prosecute them, but through subtler means. One strategy was to organize fellow planters to make numerous simultaneous calls requesting that the District Medical Officer visit the estates so that he could not appear in court to testify. Alternatively, planters would flood the hospitals with sick labourers.[206] Such sustained tactics discouraged the prosecution of planters. As shown below, the level of prosecutions under the Medical Ordinance was extremely low.

Throughout the declining years of coffee, there were persistent attempts by planters and government officials to shift the responsibility for labourer mortality from planters to *kanganies*. *Kanganies* were indeed responsible for brutal and callous treatment of labourers. Such treatment was a perfect foil for planters to argue that their only fault lay in their inability to sufficiently control the *kanganies*. This blame shifting was a form of racial displacement; *Kangany* practices, their betrayal of their fellow countrymen, were pointed to as yet another example of the low moral standards of non-Europeans.

---

203 *PPA*, 1886–87, 50.

204 Ibid., 51.

205 *SP*, 1893, iii–v.

206 Wesumperuma, *Indian Immigrant*, 263.

Table 5.1   Breaches and prosecutions in 1883 under the Medical Wants Ordinance No. 17 of 1880[207]

Clause 20 Planter Failure to:	Breaches	Prosecutions
Maintain sanitary lines	1, 120	1
Inform himself of illness and take relief measure	424	2
Send sick labourers to hospital when required by Medical Officer	49	0
Call Medical Officer in serious cases	1, 624	10
Keep a register of labourers employed and births and deaths.	7, 936	0
Clause 21 Kangany Failure to:	Breaches	Breaches
Report sickness in the gang	661	41

Source: Adapted from Wesumperuma, *Indian Immigrant*, 261.

The Principal Civil Medical Officer told the Commission reviewing the Medical Act in 1879 that 'the superintendent doesn't know much that goes on. If the cooly's sick, often the kangany will turn him off the estate and tell the supervisor that he bolted.'[208] The police became concerned about the number of sick and destitute labourers found along public roads. In June of 1878, Ameer, the police sergeant at Deltota wrote to the Superintendent of Police at Kandy, 'Sir, I beg to inform you that almost daily sick coolies belonging to different estates are found on the streets and appear a nuisance to passers-by as well as residents.'[209] This view was supported by the Annual Report on Hospitals of 1878. In Matale, for example, it was discovered that the labourers who died 'were found on enquiry to have been genuine estate coolies, who were turned away from where they were working by their *kanganies* on their getting ill and unfit to do any work, and therefore of no use to them!'[210] The Lieutenant Governor reported to the Legislative Council in 1880 that in 1878 and 1879, between 8,000 and 9,000 labourers were admitted to hospital who claimed that they had been thrown off their estates by *kanganies* because they were too ill to work.[211] However, the planters clearly turned a blind eye to *kangany* abuses when it was in the planters' financial interest to do so. Although, as shown in the table above, *kanganies* were prosecuted at a higher rate than planters, the rate was still very low. In fact, the prosecution of *kanganies* often depended on the testimony of planters who were normally reluctant to give it. As one planter said, the government

207   Ibid., 261.
208   *SP*, 1879. 70.
209   Ibid., 193.
210   *SP*, 1878, 45.
211   *CH* 1880, 54.

could hardly expect the planters to help prosecute '… our own *kanganies*—the men, practically, whom we depend on for our labour supply.'[212]

Conclusion

I have explored the struggles over the practices of authoritarian biopower during the years in which the discourse of miasmatic disease declined, but concern over sanitation gained increasing prominence. The British attempted to institute a modest programme of public health in the colony in order to bring it in line with progress in Britain. However, the social and disease environments differed between Ceylon and Britain. Furthermore, planters represented a fairly single-minded community with failing profit margins who ran an industry seen as essential to the colony. They refused to commit themselves beyond a policy of merely 'making survive' when possible and even then only when they felt they could justify the costs. Consequently, they were unwilling to do very much to safeguard the health of labourers in a life-threatening environment. State bio-politics was strongly resisted not only by planters who saw it as a new form of intrusion onto their defacto sovereignty of the estates, but also by Tamil labourers who were suspicious of European medical practices. Focusing on the examples of cholera and sanitation, I have attempted to highlight the complexity of the surveillance practiced by the state as well as the multiple forms of resistance by those who for differing reasons wished to keep the bureaucratic organisations at arm's length.

---

212 Wesumperuma, *Indian Immigrant*, 260–61.

Chapter 6

Visualizing Crime in the Coffee Districts

[T]he dream of a transparent society, visible and legible in each of its parts, the dream of there no longer existing any zones of darkness ...

Michel Foucault[1]

Introduction

Coffee estates faced a set of fundamental spatial problems. As most of the estates were hundreds of acres in extent and were surrounded by jungle, grasslands, and villages, planters could not afford to fence them. The boundaries were very porous and difficult for planters to control. In this chapter I will explore two sets of border problems; cattle trespass and coffee stealing. Meyer points out that although planters thought of the estate and the village as separate social spheres, they blurred into one another spatially and were seen by local people as having fluid boundaries.[2] There is evidence that many villagers questioned the very legitimacy of the planters' right to occupy the land. Such a failure of legitimacy stemmed in part from the fact that the Crown Lands Encroachment Ordinance of 1840 privatised what had within living memory of villagers been common lands, and in part from the fact that the Kandyan peasantry felt that the British had done little or nothing to benefit them.[3] As we have seen, porousness of boundaries and loss of control over borders between self and Other contributes to feelings of abjection adding a psychological dimension to an economic problem. The planters responded to threatened boundaries by putting in place systems of what Foucault[4] terms surveillance and Scott[5] legibility. I outline the various ways in which planters attempted to survey their extensive boundaries in the face of what they took to be a collective assault. I explore the ways estate labourers,

---

1   Foucault, 'The Eye of Power', 152.

2   Meyer, 'Enclave', 212–13.

3   K.M. De Silva, 'Resistance Movements in Nineteenth Century Sri Lanka', in *Collective Identities, Nationalisms and Protest in Modern Sri Lanka*, edited by M. Roberts. (Colombo: Marga, 1979), 139. On the other hand, Meyer points out that relative to India in the nineteenth century there was less sense of 'peasant consciousness' and hence of resistance to the British moves against their land. See E. Meyer, 'Forests, Chena Cultivation, Plantations and the Colonial State in Ceylon 1840–1940', in *Nature and the Orient: the Environmental History of South and Southeast Asia*, edited by R.H. Grove, et. al. (Delhi: Oxford University Press, 1998), 815.

4   M. Foucault, *Discipline and Punish: The Birth of the Prison* (New York: Vintage, 1979).

5   Scott, *Seeing Like a State*.

villagers and traders 'poached', to use de Certeau's[6] term, from the space of the estates. In turn, I will examine the attempts at governmentality by a state, whose first duty was clearly to capital, but was also committed to its pastoral duties to the various subject populations.

Cattle-Trespass

Cattle trespass is an important issue because it raises questions of rights to land, competing notions of what constitutes land use, the inseparability of the capitalist spheres of the plantations and peasant agriculture on small holdings, and ultimately it raises questions of the political legitimacy of British rule in Ceylon. It also illustrates the diffuseness of power and how laws set up for one purpose were used for very different ends.

The Political Ecology of Cattle

Cattle, (buffalo and black cattle), constituted an important component of the rural economy of the highlands during the nineteenth century. While there was approximately one head of cattle for every two people in the highlands, the ownership of cattle was quite concentrated, with some important headmen having hundreds of head of cattle, and poor peasants often having to hire cattle for ploughing.[7] There was also a large demand for cattle for transport of goods, especially coffee and rice to and from the estates. Even after the advent of the railways in mid-century, cattle-drawn carts, and pack animals (*tavalams*) continued to move goods short distances, and in the remote areas not served by railway.[8] With the spread of coffee estates and of peasant coffee after 1840, the amount of pasture land in the highlands decreased dramatically. Consequently, cattle were often turned loose to fend for themselves when not required for draught purposes. Such a lack of supervision meant that they were often undernourished and bred of their own accord. In addition, herds were ravaged by epidemics of murrain and hoof and mouth diseases throughout the coffee period. Both diseases were probably introduced by animals brought from south India to transport goods to estates.[9] While estates often owned draft animals, they also depended on the availability of peasant cattle for transportation off the estates.[10] As such, both the government and the planters had an interest in the well

---

6   De Certeau, *The Practice*.

7   Rogers, *Crime*, 84.

8   A.M. Ferguson, *Summary of Information Regarding Ceylon* (Colombo: A.M. and J. Ferguson, no date), estimated that in the mid-1860s there were 13,000 Sinhalese carters associated with the coffee trade. Where paths were unsuitable for carts, tavalams were used. Tavalam cattle could carry sixty to ninety pounds in sacks slung across the cattle's back. Most tavalam drivers were Moormen but some were Tamil (*ARC*, 1868, 50).

9   Rogers, *Crime*, 84–87.

10   Webb, *Tropical*, 98, states that while some estates had no herds, many kept cattle for manure and traction. The movement of goods to and from the estates was left to native carters.

being of the herds in the highlands.[11] While allowing cattle to roam was a rational response to limited pasturage on the part of the villagers, it may have been the case that indiscriminate breeding further reduced the vigour of the stock. H.S.O. Russell, the Government Agent for the Central Province made the connection between the increase in coffee estates and the decrease in the quality of cattle when he wrote in 1869 that whilst the Kandyans were unquestionably ignorant of modern principles of cattle breeding, they nevertheless had been used to keeping their cattle in herds where the strong bulls would drive out the weak and 'breeding took place on the principle of natural selection.' He went on to say that, 'now with yearly more restricted pasture, it was no longer possible to live in herds' and the breed declined.[12] But what concerned government officials, planters and villagers much more was the damage that stray cattle did to agricultural crops. As one official reported in 1867, so much damage was done to peasant gardens that 'at present there is not a single cow, or buffalo which has not done more damage to the community at large, than double her worth to the owner.'[13]

The Alienation of Common Lands

From the beginning of the coffee era there had been conflict between planters and villagers over cattle trespass.[14] The root of the problem lay in the fact that the estates occupied (*chena*) lands that had been used by peasants as common grazing lands. Even when estates were not located on traditional grazing land, they were often located adjacent to them. Under the laws of the Kandyan Kingdom, as we have seen, uncultivated lands belonged to the king with peasants having the right to use them as long as they had not been specially designated as royal forests. Peasants used these lands to collect fire wood and to supplement the rice grown on flooded valley bottoms. Unlike on their rice lands, peasants paid no taxes on the *chena* lands. In line with the teleological, modernization thesis, Bandarage[15] states that 'the clash between users' and owners' rights is an inevitable outcome in the transition from pre-capitalist to capitalist forms of land tenure.'[16] In the late 1830s European purchases

---

However, tavalams could not be used in wet weather as it was impossible to keep supplies dry (Millie, *Thirty*, Ch. 3).

11 F.C. Fisher, the Assistant Government Agent for Nuwara Eliya District reported in 1870 that murrain was killing large numbers of cattle in the district and that planters were worried about the condition of pack animals (tavalams), *ARC*, 1870, 65.

12 *ARC* 1869, 39. Roberts has argued that declines in numbers of cattle had more to do with disease than with the reduction in pasturage (M. Roberts, 'The Impact of the Waste Lands Legislation and the Growth of Plantations on the Techniques of Paddy Cultivation in British Ceylon: A Critique', *Modern Ceylon Studies* 1 (1970): 157–98).

13 *SP*, 4 1867, 155, quoted in Rogers, *Crime*, 85.

14 De Silva, *Letters on Ceylon*, 14–15.

15 Bandarage, *Colonialism*, 91.

16 Whilst there was still a great deal of common land in England at the beginning of the eighteenth century, most had been privatised by the middle of the nineteenth century (M. Watts, 'Enclosure: a Modern Spatiality of Nature', in *Envisioning Human Geography*, edited by P. Cloke, P. Crang and M. Goodwin (London: Arnold, 2004), 48–52.) The enclosure acts

of common land were often contested by the Kandyans. Governor Ward, a champion of the planting enterprise, commented that in this early period of planting:[17]

> The Government had so peacefully slumbered over its rights [to land] that twenty years after the occupation of the Kandyan provinces when European enterprise first sought in coffee cultivation a new field for the investment of capital, the local authorities were startled to find that they, literally, had not the power to give a good title, even to the most scrupulously preserved Royal Forests—grants had been made to European capitalists at an early period but every inch of land was contested by native claimants and the grantees were compelled to obtain by process of law the title which the Government was powerless to give them, or to compromise with their opponents, a proceeding which only doubled the number of frivolous claims.

In order to open up land for coffee estates, and stifle challenges to European ownership on the part of the Kandyans, Ordinance 12 of 1840, more commonly known as the Crown Lands Encroachment Ordinance, was passed. While the ordinance reaffirmed the Kandyan notion that such lands belonged to the crown, it departed from prior usage by selling off much of that common land for private use. Only if peasants could provide a deed for such lands or proof that they had paid taxes on them could they keep them as private land. But of course the British knew only too well that few could produce either document, for as common lands they were by definition subject neither to individual ownership nor to tax.

A letter to the Secretary of State for the Colonies from Governor Gregory reveals the tension between governmentality and laissez-faire liberalism:

> In these patches the natives obtain their fence sticks, wood for fuel and for building and shelter for their cattle, and it is almost tantamount to expulsion to clear them. On the other hand, their fertility makes them to be sought after by the planters.[18]

Not only did the privatization of the commons represent a 'deepening of the market as common rights were displaced by private property and land ownership and by wage work', but as Scott argues, 'it made the new capitalist order more legible by replacing the ambiguity of common land by an organisation of land that was subject to a 'standard grid...that...could be centrally recorded and monitored.'[19] It turned 'waste' land that was fiscally barren and inefficiently exploited into a positive benefit for the state. However, the peasantry largely ignored the ordinance and continued to use Crown Lands as if they were common land. Although the British sometimes removed peasants who occupied and farmed these lands, they tended to turn a blind eye to their use for the traditional purposes of gathering wood and pasturing cattle. The real conflict developed, however, when these formerly common lands were sold

---

devastated peasants who like the Kandyans depended upon common lands to supplement their crops. As Watts, (*Enclosure*, 51) writes, enclosure 'wrought a sort of confinement, dispossession, incarceration, privatisation and social transformation all at once.'

17 Governor Ward to the Secretary of State for the Colonies *CO* 54/345, No. 46 of 29 August 1859, from Turnour's letter of 15 October 1836 quoted in Enclosure No. 4, p. 173.

18 *SP*, 1873, No. 29, 9.

19 Scott, *Seeing Like a State*, 2.

to planters who then excluded peasants from using them or when cattle grazing on Crown Lands wandered into coffee estates.

This conflict proved to be not only a major source of aggravation to planters and peasants alike, but at times, such as the late 1840s, it was politically destabilising. For example, P.E. Wodehouse, a former government official and planter, reported in 1850 to the British Parliamentary Commission inquiring into peasant unrest in 1848 that:

> it stands to reason that all the views of a European settler should be at variance with [those of the villagers]. For example, his estate may be surrounded with land which they have for generations looked on as common property; he gets into disputes with them immediately respecting the trespass of their cattle and it is almost impossible to define any fair line between the rights of natives in respect of that ground and the rights of Europeans whose cultivation is injured.[20]

In some places there was local resistance to what was seen as the expropriation of common land by planters. The coffee planter George Ackland in testifying before the British Parliamentary Commission inquiring into the unrest of 1848 said:

> Upon one occasion when I first settled in the Valley of Doombera where the natives are very superstitious, and generally speaking troublesome to deal with, they threatened to murder my people; they burnt down a new house that was built; the headman (native chief) himself told me that he would never allow us to settle there. This was the first plantation that was made in the interior upon any extensive scale. They also complained that their grazing lands were claimed by the government, and that they attributed to the coffee planters making their applications to purchase land; therefore upon these grounds they were dissatisfied.[21]

Another planter spoke of a confrontation while planning a road to his new coffee estate in 1844,

> I entered into conversation with the spokesman who was an *aratchi* [village headman] and told him the advantages the road would be to himself and the surrounding villagers. He became very excited and in a very insolent manner said "who sent you white people here, we did very well without you; look there" (pointing to a coffee estate) "that forest was mine"; and then to a sugar plantation "that was mine"; then to open ground upon which some estate cattle were grazing "that was mine"; you have levelled our forests; seized our chenas and now you are turning our paddy fields into roads. But we have a man up there (pointing to the Knuckles range of the mountains about 12 miles distant) who will soon get rid of you; he will cut the ... (drawing his hand across his throat with a vicious smile) of everyone of you ... This statement of the *aratchi* betrayed the voice and feelings of the Kandyans in general.[22]

---

20 *British Parliamentary Papers (BPP)* 1850, Volume 12, 140.

21 Ibid., 19.

22 Colonel Watson, quoted in R. Pieris, 'Society and Ideology in Ceylon During a Time of Troubles, 1795–1850, Part 2', *University of Ceylon Review* 9 (1951), 279.

It was not uncommon for peasants in the 1840s to resist the estates' invasion of their common lands by clandestinely tearing down estate boundary fences and chopping down coffee trees. This became such a problem for planters that Ordinance 6 of 1846 was passed to prevent the injury or theft of estate products. Clause 14 of this ordinance stated that anyone who 'unlawfully or maliciously cut, broke, barked or uprooted or otherwise destroyed trees on a plantation could be imprisoned with or without hard labour for any period not exceeding one year, and if a male, to corporal punishment not exceeding fifty lashes.'[23]

The problem of planters buying traditional grazing lands continued throughout the coffee period. F.C. Fisher, the Assistant Government Agent for Nuwara Eliya District reported in 1870 that the inhabitants still complained of being deprived of their grazing lands and their buffalo been shot. The AGA reported that such conflicts could not be avoided and that the remedy was for villagers to experiment with growing pasture near their villages.[24] It is telling that the AGA should argue for this solution rather than suggesting that planters should fence their land. The plan to have villagers grow pasture near villages had, from the point of view of government, the dual advantage of reducing trespass and allowing the supervision and hence improvement of stock. However, this plan was not taken up by villagers for two very good reasons. Firstly grazing land was so unproductive that an untenably large area would have had to be fenced by villagers. And secondly, villagers could not have done so because the land was not theirs to fence. By the following year the Government Agent of the Central Province, H.S.O. Russell became so concerned by the starvation of cattle from lack of pasture that he wrote,[25]

> I am endeavouring now to prevent the sale of common pasture land… but the evil of indiscriminate alienation of such land has already been carried so far that in many places no pasturage is left for villager's cattle, and in others beasts can be turned out to graze only at the risk of their straying into neighbouring coffee estates, where they will be impounded or shot.

The Framing of the Cattle Trespass Ordinances

In light of the government's unwillingness to make planters fence their estates, the only alternative seemed to be to establish a set of penalties for trespass and hope that peasants would be forced to adopt different herding practices. From the government's point of view, such a policy, rather than favouring planters, was seen as a rational solution for the benefit of private property and the state. As we have seen, the government generally did not approve of the nomadic grazing practices employed by the peasantry. Such practices were regarded as irrational and premodern in that they damaged the breeding stock and private property alike. Just as the state wanted a settled, rationally controlled and legible population, so it wanted the same for the herds of cattle.

---

23 Vanden Driesen, 'Land sales 2', 42.
24 *ARC*, 1870, 65.
25 *ARC*, 1871, 41.

The first legislation was the Cattle Trespass Ordinance No. 2 of 1835, which empowered the owner of a property to seize cattle found on his property and hold them until the owner was located and the damage assessed. Notice of seizure was supposed to be reported to the nearest constable, *vidahn* or local headman,[26] if one was to be found within 80 miles. Otherwise no claim could be made. The proprietor could in theory recover the full costs of the damage by bringing a civil action in the Court of Requests. If the trespass was proved to have taken place at night the owner of the cattle was required to pay a fine of double the value of the damage done. Cattle that could not be caught could be shot. But before doing so, the proprietor had to swear an affidavit before a district judge or magistrate stating that the cattle habitually trespassed and could not be caught. Such licences were valid for a month.[27]

The government encountered a series of difficulties in administering the 1835 ordinance. These stemmed from the perennial lack of a sufficiently dense network of administrators on the ground to implement ambitious bureaucratic solutions. This, in turn, was due to a series of interlocked decisions between London and Ceylon about general government spending priorities and the choice to invest most heavily in infrastructure including roads and the railways and as a result operate with only a very sparse police force and judiciary.[28] These financial compromises produced various problems: The ordinance was impossibly bureaucratic and ill-adapted to the lightly populated rural areas where the offences took place. The injured party might need to travel up to 80 miles over tracks and dirt roads to report the trespass and might then very likely be faced with a Headman who was unfamiliar with the bureaucratic procedures. B.F. Hartshorne, the AGA. of Nuwara Eliya summarized the problem well in 1873 when he wrote,[29]

> In order that a man may recover the damages to which he is entitled under this Ordinance, it is necessary for him to produce a report from the nearest headman, drawn up with several precise formalities, which are so numerous and so easily misunderstood or omitted by the headmen that the report is invariably informal, and the complainant can consequently recover no more than five shillings a head in respect of the cattle which may have committed damage valued at five pounds. ...I have heard hundreds of cases under this Ordinance, and never yet seen one report which was in conformity with its requirements.

A second problem was structural and affected the implementation of many ordinances. As a cost-cutting measure, the state appointed European planters in remote rural areas to serve as magistrates. Thus the magistrates who judged the merits of a case

---

26 Policing in rural areas was overwhelmingly conducted by unpaid village headmen.

27 Cattle Trespass, *The Ceylon Civil Service Manual* (Colombo: F. Fonseka, Printer, 1872), 126.

28 Legg notes that the police in India relied much less upon surveillance than their counterparts in Britain. Rather they were more apt to rely on force and violence as they assumed that lack of self-discipline amongst non-Europeans demanded a greater level of violence. While such a view would have been held by police officials in Ceylon as well, the major reason for the relative lack of surveillance was financial. (see Legg, *Spaces of Colonialism*, Ch. 3).

29 *ARC*, 1873, 53.

often ruled on cases of trespass on, either their own, or their friends' properties.[30] The following is an interesting example drawn from the reminiscences of a planter: 'we were tormented by buffalo trespass [and this created] very bitter feelings between the natives and the Europeans.' He went on to describe a case of trespass on his land and how he had used the villagers' buffalo for days for draught purposes on the estate even though the owners offered him a pound per head to return them. When the owners finally threatened to take him to court he reminded them that he was the magistrate who would decide the case. Finally he returned their buffalo when he discovered that it was another villager who had a grudge against the owners of the buffalo who had driven them onto his estate in order to cause them trouble.[31] In the early years especially, villagers had very little recourse if planters impounded or shot their cattle. Planters widely ignored the requirement that permission be acquired from a court before shooting cattle and more often than not assumed the sovereign right to deal as they saw fit with animals that came on their land. Such rights, although not legally granted, were tacitly acknowledged.

The fact that villagers sometimes drove cattle onto an estate to settle a grudge with another villager reveals how bureaucratic rules can escape their originally intended purposes and how diffuse and fragmented power can be. Estate labourers also sometimes profited from cattle trespass laws as evidenced by the following report by H.S.O. Russell, the GA for the Central Province in 1869:[32]

> The chief complaints which have been made to me by villagers when I was travelling through the District have had reference to the impounding and shooting of their cattle when trespassing on coffee plantations. They allege the Tamil coolies not infrequently drive cattle on to an estate, and then tie them up, and either extort a large fine for the release of the impounded animals or slaughter them and use their flesh for food.

While cattle trespass was a source of much social conflict, there were, as we have seen, individuals who could turn that conflict to their benefit. In this way, villagers could exact revenge on fellow villagers and poor labourers could secure food or supplement their low wages at the expense of villagers.

Dissatisfaction with the old ordinance led to the framing a new ordinance. Ordinance 9 of 1876 was passed allowing landowners to confiscate cattle which had strayed onto cultivated or fenced land and turn them over to the local village Headman. The owner of the cattle was then held liable for damages. Although the new law cut down on travel to report trespass, the shifting of power to local headmen created new problems. It was commonly reported that unscrupulous peasant landowners colluded

---

30 In most rural areas, as a cost-cutting measure, the magistrate and Justice of the Peace was the same person. A Justice of the Peace could only take evidence and refer the case to the Queen's Advocate. Magistrates tried cases in Police Court and were able to set maximum punishments of three months in jail, twenty lashes and five pounds fine, or any two of the three. District Courts could give a maximum of one year in jail, fifty lashes, one hundred pounds in fine or any two of the above. The Supreme Court had no limit on the nature of its sentences (Rogers, *Crime*, 44).

31 Anonymous, *Days of Old*, 14.

32 *ARC*, 1869, 39.

with corrupt headmen to drive stray cattle onto their own land and get damages assessed at an inflated rate.[33] The various cattle ordinances were never able to greatly reduce the social conflict around cattle trespass. In fact, as we have seen, they often generated conflict. Furthermore, there were relatively few convictions. Figures from the Police Courts show that between the inception of the Cattle Trespass Ordinance of 1876 and 30th September 1880, of 3,141 cases instituted, only 266 convictions were obtained.[34] Consequently, alternate forms of sanction were arrived at. Cattle were killed or used for a time for draft or a bribe was paid by the owner to secure their return. Because of the problems of the ordinance's implementation, in 1886 it was proposed in Legislative Council that the Cattle Stealing Ordinance of 1876 be amended to require estates to be fenced if they were to claim more than a nominal one rupee damage for cattle trespass.[35] In proposing this amendment the Acting Attorney General argued that,

> it has been found that the Ordinance bears rather harshly on village communities. I think it is but fair, that, while subjecting the inhabitants of native villages to the heavy penalties prescribed by the Cattle Trespass Ordinance, those who deliberately open estates in the midst of such villages should in turn undertake all the responsibilities which fairly attach to their position. No one need wonder if cattle do stray from a public pasture ground or a private field into a tea estate if there is no barrier between them.[36]

Predictably, the planters reacted very negatively to the proposal arguing that this would encourage peasants to neglect their cattle and that fencing estates was ruinously expensive. They bemoaned the fact that the Government appeared to no longer support their interests. As one planter put it, 'what does the cattle-owner do for the Government? [Whereas] the owner who cultivates his land ultimately contributes towards the increase of revenue and therefore is entitled to greater consideration.'[37]

The 1886 amendment requiring the fencing of estates reversed a half-century long of cattle trespass policy. It was a tacit admission by the government that it was unable to enforce the law against cattle trespass. The government was frustrated by the nomadic character of cattle and the fact that they could not easily be traced to owners. They were also frustrated by the lack of surveillance of the boundaries of the estates and the resulting disputes over whether cattle had wandered in on their own or had been driven in by villagers or by Tamil labourers. As rural courts were overburdened in large part by the peasants who flooded them with disputes with other villagers, the authorities were reluctant to deal with trespass cases for they were known to be unlikely to produce convictions. The illegibility of cattle, failure of the legitimacy of the ordinance in the eyes of villagers and estate labourers, and the lack of funds to properly survey and punish, left the boundaries between the estates and the villagers nearly as porous as ever.

---

33 Crawford, in *ARC* 1902, G8, cited in Rogers, *Crime*, 105.
34 *SP*, 1881, 33.
35 Throughout the coffee period a pound sterling was worth approximately 10 rupees.
36 *CH*, 1885–86, 94.
37 *PPA*, 1886–87, 15.

154   *In the Shadows of the Tropics*

Theft of Coffee

Throughout the coffee period planters worried about how to stop the theft of coffee. The estate labourers regularly pilfered from the fields and store rooms, Sinhalese villagers raided the estates at night, while cart men and boat men stole coffee after it left the estate. In this section, I will examine the tactics used in coffee stealing and the strategies of surveillance developed by planters and authorities in their attempts to counter theft. As was the case with cattle trespass, the lack of a reliable network of surveillance within and along the borders of the estates, made theft difficult to prevent.

On The Illegibility of Coffee and its Consequences

Throughout the coffee period planters knew that they were losing a modest amount of coffee en route to market. The difficulty in preventing theft was exacerbated by the sheer number of potential sources of theft and the lack of surveillance of the roads and waterways between the estates and Colombo. To give an example, in 1873 a planter sent 300 bushels of coffee to Colombo by cart and rail. At Colombo it was measured 3 bushels short. He wrote that it was

> either stolen by the cartmen, who have in that case to bribe the storekeeper to give a full receipt, or it is stolen by the storekeeper himself, or by the railway officials, or by the cartmen or boatmen who deliver it to the stores. There is no check on anyone from the time it leaves the estate till it is received in Colombo.[38]

When coffee remained in the hands of a single shipper, theft was more difficult, and yet even then it was routinely stolen.

Millie[39] writes of the tricks employed by carters to steal coffee. One was to claim that they had been given less coffee than they had in fact been given. He says that this was hard to combat because carters always claim that the load is short 'no matter how it is measured, because they always steal some.' Such a strategy worked because virtually all carters employed it. As the value of coffee rose on the international market, thieves became bolder and whole cart loads disappeared. The government responded by registering carts, but this was no guarantee as some carters used forged plates so that stolen loads of coffee could not be traced.[40] In 1879 a more systematic plan for registering and recording the load on carts was put in place. Carts were to be checked in at police stations along the road each day to ascertain that the load had not been tampered with. The daily information was then forwarded to the next station and to Colombo so that the progress of each cart could be followed. As the Inspector General of Police wrote, 'each cart was thus kept in sight from day to day. Formerly it was lost sight of for weeks.'[41] But such a strategy for making coffee more visible could not combat the carters and *tavalam* drivers' tactic of pulling over into

---

38 F.W. Byrde, 'Correspondence on the subject of coffee stealing', *SP*, 1873, 172.
39 Millie, *Thirty*, Ch. 7.
40 *PPA*, 1864–65, 27.
41 *ARC*, 1883, 29c.

a secluded area, removing a portion of the coffee from the sacks and making up the weight by a 'judicious damping of the beans.'[42]

Coffee was also moved by canal boat between Ratnapura and Colombo and here there were systematic losses as well. The Inspector General of Police in 1881 commented that, 'from enquiries made I have learnt that small fortunes have been made for many years past by the coffee stolen from cargoes on the way from Ratnapura to the Colombo stores, and that the quantity stolen has been made up by wetting or mixing; but until lately no complaints have been made [because the thefts remained undetected].' He went on to describe the ways in which coffee was stolen. One way was by mixing, a small amount of inferior coffee with the high quality coffee. More common was wetting, a more complex but less expensive system:

> The plan until lately was, to remove two or three measures and sink the sack of coffee in two or three feet of water, where there was gravel or grass, so that the sack should not be soiled, leave it there for ten or fifteen minutes, and then dry it on a rock in the hot sun; but now the boatmen are more cautious [due to increasing surveillance]. They run the boats into one of the side ellas or canals, and the coffee is there emptied from the sacks on to a carpet of sacking in the bow of the boat, and after it is well mixed with water which is thrown on it from a pot, it is put back into the original dry sack and sewn up. The quantity removed from each sack before the wetting commences depends on the quality of the coffee. If of the best quality, three measures are taken out, but from the worst quality one and a half measure only.; because of the inferior qualities will not swell. The parchment does not begin to swell until the third day after wetting and on the sixth day it is at its full size for delivery at the Colombo stores. After the sixth day it begins to shrink, and will go on shrinking until it becomes less than it was originally. So that the boatmen will always try to deliver the coffee on the sixth day after wetting.[43]

The brilliance of this system was that on its arrival in the warehouses in Colombo the coffee measured correctly to the specified paperwork both in volume and weight. It was only later that the discrepancy became evident and by then it appeared to have been stolen from the warehouse. This system of wetting had to be stopped, not only because a great deal of coffee was stolen in this manner but more importantly because wetting reduced the quality of the remaining coffee and hence its price on the London market. As with the carters, the only way that it could be stopped was through making the coffee visible throughout its whole journey by boat. Consequently, boats were placed under police surveillance, with constables at posts every sixteen miles, constantly patrolling eight mile stretches on either side of the post.[44] While this plan appeared to meet with some success, it was difficult to judge whether the carters and boat men could have countered this surveillance strategy and found other ways to poach. Unfortunately for the planters, the villagers, the carters and the boatmen, time ran out for them all and within a few years coffee was dead in the highlands.

Planters lost even more coffee to estate labourers who exchanged it for goods at bazaars. As the price of coffee doubled in the early 1870s, planters began to notice a dramatic increase in thefts from estates. During the 1870s such thefts greatly

---

42 *ARC*, 1868, 50.

43 *SP*, 1881, 16–17.

44 *SP*, 1881, 17.

exceeded losses through shipping. The problem with coffee beans is that they all look pretty much alike. They are an illegible good. Planters knew that they were losing a lot of beans, but as villagers also grew coffee, the planters could not prove their ownership.

The Planters' Association began to be concerned in the mid-1860s when large numbers of pulpers and boutiques dealing in coffee began springing up around estates at harvest time. Given the increase in theft of coffee, they thought that the 5 shilling fine for theft was an insufficient deterrent.[45] As they could not persuade the government to ban these establishments, they tried to convince the government to make it difficult for them to procure and sell coffee. To this end, the Planters' Association argued that 'natives' with coffee should be made to account for where it came from. Pressure mounted on the government to act throughout the late 1860s and early 1870s as planters relentlessly campaigned to have the law toughened.[46]

By the early 1870s government officials were becoming convinced that a new law was in fact necessary. Framing such a law was problematic because peasants also grew a lot of coffee. The planters' evaluation of the problem of coffee stealing was premised upon racist beliefs about native character and rights. The case for a new law and the problems of enforcement were summed up by the B.F. Hartshorne, the AGA for Nuwara Eliya in 1872:

> The complaints of coffee stealing have latterly been very numerous; but cases are rarely and with great difficulty proved. The Tamil coolies on the estates are constantly pilfering small quantities, whilst low-countrymen invade plantations in a body by night with sacks, and strip the trees of their leaves as well as the ripe and unripe coffee altogether. The Moormen seem to be the chief receivers; and the Kandyan Sinhalese are less frequently engaged in these nefarious practices than any other class of persons. Amongst the numerous cases that have come to my knowledge, I do not remember a single instance in which a Kandyan was caught stealing, otherwise than under the wing of a low-countryman, Moorman or Tamil man. ... I would recommend that no person be allowed to deal in coffee, except for consumption upon his own premises, without a license.... Managers and owners of estates of more than fifty or a hundred acres might be exempted, and the possession of any coffee by any person not being able to account for it satisfactorily should be declared an offence.[47]

In 1873 the Coffee Stealing Bill came before the Legislative Assembly. The Governor Sir William Gregory in a letter to the Chief Justice laid out what he saw as the problem and sought a proposed law.[48]

> If the accounts which I receive be correct, it has become the custom for persons to settle themselves in the vicinity of an estate for the express purpose of trafficking in stolen

---

45 *PPA*, 1864–65, 6, 7.

46 *PPA*, 1872–73, x. In fact, police figures for 1872–73 demonstrate that while relatively few convictions were obtained and fines were in the range of five to ten rupees, the fines were normally accompanied by a sentence of one to three months hard labour and in a number of cases five to twenty lashes (*SP*, 1873, 154–59).

47 *ARC*, 1872, 1873, 71.

48 Gregory to Chief Justice 10 January 1873 in *SP*, 1873, 163.

coffee. It seems there is great difficulty in apprehending the offenders, they being, in too many cases, coolies employed either on the plundered estate or in the neighbourhood, and that the only way to meet the evil is to attack the real instigators of all this plundering— namely the receivers

Some time ago, when sheep stealing was rife in Ireland, an Act was passed placing the onus on every one in possession of mutton of proving that it was legitimately obtained. The same legislation was applied to timber, and it has had a remarkable effect in putting down sheep and timber stealing. Could not some such principle be applied to coffee, and persons found in possession of it be compelled, if called on, to prove that they obtained it honestly?

The Chief Justice in his reply to the Governor proposed a framework of law based upon two earlier ordinances against cattle stealing. He argued that this was appropriate as it was unwise to institute a law that appeared to be solely in the interest of the coffee planters. At the heart of the proposed law were the following clauses: that it was illegal to possess more than two measures of coffee whose provenance could not be accounted for; that any coffee of more than a bushel that was not purchased from a warehouse or merchant must have a note of sale describing the amount and kind of coffee, any marks on the bag and the name and address of the seller; that any bill of sale had to be attested by a member of the police or Headman of the district; and that the police had power to search a suspect's house without warrant and to provide stiffer punishment for habitual offenders.[49]

The government was well aware that this legislation would, perhaps unnecessarily, increase bureaucratic involvement in trade. Even more problematically it placed the burden of proof of innocence on the peasantry. This was evidenced by a letter from the Colonial Secretary to the Secretary of the Planters' Association expressing the Governor's concern that unless legislation was drafted very carefully, 'great hardship is sure to be inflicted on the innocent native trader—in fact, the total destruction of an active and profitable traffic might be caused by the imposition of new and vexatious restrictions on the transfer of coffee.' The Colonial Secretary then suggested that the Planters' Association collect for the government 'accurate reports as to the prevalence of this crime in the different plantation districts, by whom the thefts are committed, whether by natives or by estate coolies, and who are the receivers.'[50] It is a mark both of the influence of the planters and the weakness of the state bureaucracy that the government should ask the Planters' Association to collect information on the provenance of theft. The request reveals just how little surveillance the government had in the coffee districts.

Several months later the Secretary of the Planters' Association replied to the Colonial Secretary informing him that the Association had received reports from 133 estates summarised as follows. Thefts were prevalent in 15 of the 18 planting districts and while the chief depredators were estate labourers who had 'many facilities for securing and secreting coffee', thefts were also committed by Kandyan

---

49 E.S. Creasy, Chief Justice to W.H. Gregory, The Governor, in *SP*, 1873, 163–64.

50 Henry Irving, Colonial Secretary to H. Byrde, Secretary to the Planters' Association, *SP*, 1873, 165.

villagers, low-country Sinhalese, cart drivers, and tavalam drivers. The planters identified the chief receivers of stolen coffee as 'owners of small properties near estates (Sinhalese and Tamils) Moormen traders and boutique and tavern keepers.'[51] Clearly the planters thought that men from every group was either stealing coffee or receiving it. The planters' survey was a strategic document making a case for a strong law to deal with what they argued was a systematic theft from a small number of Europeans by the non-European majority.

Surveying the Theft of Coffee: The View from the Estate

The planters' reports provide insight into their racialized understandings of non-Europeans.[52] The estate labourers were thought to be like rather weak-minded, naughty children who could and were led astray by other groups. While the labourers could not be trusted and required discipline, they were considered only partially responsible for their own behaviour. The Kandyan villagers were seen as less simple-minded than the Tamil labourers and hence much more culpable. However, they too were often spoken of as falling under the baleful influence of other more devious racial groups. Low country Sinhalese, *chetties* (Indian money lenders) and especially the Moors were seen as the malign influence behind the problem.[53]

The theft of coffee was difficult to prevent for a variety of reasons. First, coffee could be stolen in very small quantities, a few handfuls at a time, and disposed of in bazaars in exchange for food or drink. Second, the estates were large and their boundaries unprotected and therefore coffee could be stripped from remote fields at night or hidden in jungle on the periphery of the estate to be collected later. Third, as coffee beans look alike, it was virtually impossible for planters to identify their own beans in village boutiques. And fourth, planters sold their own lower grade coffee to villagers for local consumption and therefore there was estate coffee being sold legally in boutiques. It was within this context of illegibility that coffee stealing flourished.

The following are some of the tactics adopted by estate labourers. Just as the labourers were watched, so they in turn used their superior numbers to watch the planters. As one planter said, 'it is almost, I think, impossible for a superintendent in charge of 300 or 400 coolies—as he may be watched on every side, and advantage taken of all his movements to detect theft, except in very few cases.'[54] Labourers often used sickness to steal coffee. One planter wrote:

> this year I caught two of my own coolies, who had given their names in the morning as being sick, picking coffee at least two miles from the place where I was then picking with my whole force, and not so far from the village where a low country man had built a shed,

---

51 H. Byrde, Secretary of the Planters' Association to the Colonial Secretary, 19 May 1873 *SP*, 1873, 166–67.

52 'Correspondence on the subject of coffee stealing', *SP*, 1873, 163–217.

53 The issue of colonial images of the differing degree of criminality among races was discussed in Chapter 1.

54 P. MacRae, 'Correspondence on the subject of coffee stealing', *SP*, 1873, 202.

and exposed fish and coconuts for sale; and it was quite clear to me that the coffee was never intended to be brought to my pulping house.[55]

Another strategy was to have 'small quantities of coffee pilfered by little children, and taken down to bazaar by the men on Sunday.'[56] The most common place to keep stolen coffee was 'in the lines, where it is evidently being kept till a convenient opportunity occurred to carry it away. I have had a cooly arrested on his way to the bazaar, who had converted the skirt of his coat into one large pocket, in which was found a quarter of a bushel of parchment coffee.'[57]

While labourers were thought to steal the most coffee, villagers were thought to do the most damage as they usually raided the plantations at night breaking branches and stripping unripe coffee along with the ripe. Planters wrote of finding whole fields of stripped and broken trees, with ripe and green coffee scattered all over the ground. Some thieves were bold enough to do this during the day. As one planter wrote:

> [o]n Appogastenna Estate, Kegalla, two years ago, armed gangs of twenty to twenty five men—Sinhalese villagers—entered the estate, drove away the watchman, and picked a considerable quantity of coffee off the trees before he could run down to the estate and give the alarm and get assistance.[58]

Villagers were also thought to collude with labourers. One planter described such a case as follows, the 'store was entered at night by Sinhalese and six bags removed in one night; robbery was probably connived at by estate coolies.'[59]

One of the principal strategies used to protect the estates was the posting of watchmen in remote fields, on the estate's borders, and particularly at the storehouse where picked coffee and tools were kept. But as many planters noted, the watchmen themselves stole or turned a blind eye to theft. Planters complained of often being unable to catch thieves even when they saw them, and consequently appealed to the government to be allowed to 'shoot with small shot any person seen carrying away coffee from a store or estate without permission.'[60] Another said, 'A license should be given to every proprietor of fifty acres' cultivated land and upwards to shoot thieves as he would cattle, although not in such a deadly way.'[61] The planter James Taylor wrote that it was very difficult to catch thieves from the villages because:

> there are but one or two Europeans on an estate, who can be easily watched, and are watched, when stealing is going on. Our coolies may catch them sometimes, but what then if coolie evidence is disbelieved in court? I have so often found coolies tell lies against each other and falsely blame each other – that I would very readily doubt their evidence myself.[62]

---

55 J. Gidlow, 'Correspondence on the subject of coffee stealing', *SP*, 1873, 188.
56 J.J. Garioch, 'Correspondence on the subject of coffee stealing', *SP*, 1873, 169.
57 D. Carson, 'Correspondence on the subject of coffee stealing', *SP*, 1873, 190.
58 A. Brown, 'Correspondence on the subject of coffee stealing', *SP*, 1873, 214.
59 A.C. Hoare, 'Correspondence on the subject of coffee stealing', *SP*, 1873, 190.
60 H.P. Walker, 'Correspondence on the subject of coffee stealing', *SP*, 1873, 168.
61 A. Brown, 'Correspondence on the subject of coffee stealing', *SP*, 1873, 214.
62 J. Taylor, 'Correspondence on the subject of coffee stealing', *SP*, 1873, 185.

As I have suggested, it was difficult to identify stolen coffee once it was off the estate. The borders of estates were full of 'low-country Sinhalese who buy or rent a small piece of land near estate roads, plant a few coffee trees and never weed it, but still manage to send large quantities of coffee [to market].'[63] Planters also wrote of people who formerly worked on the estates setting up boutiques during harvest time to receive stolen coffee. These included former estate labourers, as well as Sinhalese carpenters and others associated with the estates. It was feared that they used their contacts with labourers to systematically rob the estate. C.E. Kay wrote of finding 'a small Chetty near the estate [that] had bribed some of the estate coolies to steal the coffee during the night and deliver the same to him, for which he paid cash.'[64] One planter noted a strategy used by merchants in his district to make stolen coffee even more difficult to detect. The merchants:

> go about amongst the estates, buying up all sorts of rubbish [coffee] at fabulous prices, willing in fact to give almost any price to get it to mix with stolen coffee; and when any planter loses a bushel or two when dispatching, or may have suspicion that such a one is buying coffee from his coolies, he goes and gets a search warrant from the Justice of the Peace, and finds perhaps ten, twenty or thirty bushels of rice or clean coffee in the man's house, who replies to questions, that he bought so many bushels *poke* coffee at such and such an estate, that Mr. So-and-so sold him so many bushels second parchment.[65]

The planters clearly believed themselves to be under siege and felt that no non-Europeans either on the estate or off could be trusted. As one planter put it:

> Native traders of every description—from the man who bakes hoppers and sells plantains, to the butcher who supplies us with beef, the Chetty who sells rice, the cattle-shed keeper, the man of all trades, the man of no trade, the toper, the swell shopman who deals in brandy and biscuits—all are ever ready when we begin crop to receive coffee from our coolies, principally for their different wares.[66]

Planters believed that the present ordinance was not up to the task of dealing with the organised theft of as elusive a commodity as coffee, especially as there was no police presence in the coffee districts. Consequently planters often extended their surveillance beyond the boundaries of the estate. As one said, 'I have a Sinhalese watchman on the road, who has caught a great number of coolies with stolen coffee; this man informs me the Udispattu bazaar-keepers are so angry with him that he is afraid to go past their bazaars.'[67] Another planter said, that while watchmen were useful, 'spies on the boutique keepers, who annually establish themselves in the neighbourhood in crop time, have been attended with better results than that of watchmen.'[68] Planters were unhappy about the need for warrants to search village

---

63 P. MacRae, 'Correspondence on the subject of coffee stealing', *SP*, 1873, 202.

64 C.E. Kay, 'Correspondence on the subject of coffee stealing', *SP*, 1873, 201.

65 G. Shand, 'Correspondence on the subject of coffee stealing', *SP*, 1873, 175. Poke coffee was low grade and considered unfit for the international market.

66 G. Shand, 'Correspondence on the subject of coffee stealing', *SP*, 1873, 175.

67 C. Young, 'Correspondence on the subject of coffee stealing', *SP*, 1873, 189.

68 D. Mitchell, 'Correspondence on the subject of coffee stealing', *SP*, 1873, 213.

Figure 6.1   The bazaar

Source: Hamilton and Fasson, *Scenes in Ceylon*, 1881.

boutiques and houses. Rather it was argued that the manager of an estate should 'have the power to search houses without the aid of constables—merely taking another European or his own servant with him. For by the time a constable was secured, all of the village would be on the alert, in which case they are worse than useless.'[69]

Because of the lack of police presence and the difficulty in identifying stolen coffee, there were relatively few convictions. While there were a good number of planters who were Justices of the Peace authorized to charge people with theft, the magistrates often dismissed the charges for lack of evidence.[70] This outraged planters who appeared to care little for the 'hard evidence' demanded by magistrates. They argued that circumstantial evidence should be sufficient to convict. As one planter put it, '[t]he cause of failure [of most cases] was on account of insufficient evidence, as in these cases so much evidence is required to convict the thieves, unless they are actually caught in the act; estate parchment being all so much of a sameness, that you cannot swear to your own property.'[71] Another said,

> [t]here are few planters who think it of any use taking cases to Court, for it is next to impossible to get a conviction. The most of the Magistrates are mere boys, without legal knowledge or experience of any kind, and if they do convict, the punishment is so inadequate that it encourages rather than deters coffee stealers. Therefore [I] prefer making the parties pay for stolen property; but this cannot be always done, as the coolies are often or always in debt; and to turn them off the estate is only depriving yourself of their labour, without being any punishment, for they can find ready employment on a

---

69 A.H. Thomas, 'Correspondence on the subject of coffee stealing', *SP*, 1873, 173.

70 It was noted earlier that Justices of the Peace can only collect evidence for the Queen's Advocate.

71 W. Bisset, 'Correspondence on the subject of coffee stealing', *SP*, 1873, 172.

neighbouring estate, where possibly they may succeed better in thieving without being detected.[72]

Some planters believed that the magistrates were biased against them: 'It is a very good axiom that a prisoner should be looked upon as innocent till he is proved guilty; but judging from my own experience, ...and what I have read recently in the papers, the Magistrate, as a rule, looks upon the planter as the guilty party, and himself as the protector of the poor oppressed native.'[73] The cost of the proceedings was also seen to outweigh the benefits. As one planter said, he 'succeeded in obtaining a conviction at the cost of much time and money, say about $50 for only one bag of coffee.'[74] The issue of coffee stealing further increased planter impatience with what they saw as government intervention in issues concerned with running of estates. Rather the planters would have preferred that the state grant them semi-sovereignty in protecting their interests which they saw as ultimately being coincident with the state's interests.

In response to the question on a survey enquiring as to what punishment should be meted out for theft of coffee, planters called for flogging followed by long sentences of hard labour. For villagers, it was felt that a suitable punishment was 'giving lashes, to be inflicted in the convicted man's village, and before the whole of the villagers.'[75] For estate labourers many thought that 'a severe flogging on the estate barbecue [where coffee is dried] in presence of the other labourers would act as a warning.'[76] There was a tension between the government and the planters on the issue of the harshness of punishment. While planters called for harsher punishment, the government tried to move away from corporal punishment.[77] As Collingham points out, by mid-century civil servants in South Asia began to see planters and the military as violent, rather unsavoury social types who had to be controlled so that they would not do more harm than good to the imperial project. During the early 1870s the government discontinued the practice of Police Magistrates being allowed to order flogging. Such power was now to be reserved for higher courts. This move was deplored by the planters who argued that nothing less than flogging would serve as a deterrent. As the following demonstrates, some planters called for

---

72 A.F. Harper, 'Correspondence on the subject of coffee stealing', *SP*, 1873, 174.

73 J. Ryan, 'Correspondence on the subject of coffee stealing', *SP*, 1873, 169. Most judges in mid-nineteenth century Ceylon were civil servants with no training in law. It was argued by the Ceylon Government that knowledge of the people was more important than legal training, (V. Samaraweera, 'British Justice and the "Oriental Peasantry"; the Working of the Colonial Legal System in Nineteenth Century Sri Lanka', in *British Imperial Policy in India and Sri Lanka, 1858–1912*, edited by R.I. Crane and G. Barrier. (Columbia: South Asia Books, 1981), 112). It is likely that some of these civil servants were prejudiced against planters whom they saw as a rough, uncivilized lot.

74 J.C. Albrecht, 'Correspondence on the subject of coffee stealing', *SP*, 1873, 212.

75 A. Sparkes, 'Correspondence on the subject of coffee stealing', *SP*, 1873, 181.

76 J.A. Spence, 'Correspondence on the subject of coffee stealing', *SP*, 1873, 186.

77 Public flogging was abolished in Britain in 1862, as it was thought to arouse dangerous passions in crowds and hence be a threat to the civilizing process (Wiener, *Reconstructing the Criminal*, 93–94).

extraordinarily harsh punishment for theft of what was often less than a bushel of beans; they also suggested different punishments for different races. The fiercest desire for retribution was directed at the Moors and lowland Sinhalese, even though the largest losses were to theft by estate labourers. As one particularly harsh planter wrote:

> I think any native found in the act of stealing coffee, in any shape, from an estate, ought to be transported, or imprisoned for life with hard labour. When estate coolies are convicted, I think they ought to be thrashed on the estate from which they stole the coffee...and then put in jail for a longer or shorter time, owing to the gravity of the case.[78]

However, not all planters were of one mind; some worried about courts passing very harsh sentences, given what they saw as the incompatibility of 'native character' and the British court system. As one put it:

> [t]he administration of justice is so very difficult amongst Orientals, seeing they consider their oaths and lives of so little importance, that a mere matter of malice might end in the transportation (for coffee stealing) of the weakest, and most likely the most honourable of the two parties.[79]

Planters and government officials alike despaired at what they took to be widespread perjury in courts. Such was the case not only when non-Europeans contested cases against Europeans, but also against each other. To the British, this was clear evidence of moral failure and a justification for British rule. It also was held to be a powerful argument for non-Europeans not to be treated as juridical subjects but as subjects of planter sovereignty who could receive summary justice on the estates. Rogers, however, offers another explanation.[80] He says that the modern European notion of truthfulness of testimony in legal proceedings was not shared by the Sinhalese or Tamils. Rather than equality of all individuals before the law, locals believed that other factors such as the social position or known character of the defendant and plaintiff should take precedence. Truthfulness of testimony was seen as secondary to these higher values. Justice, as they understood it, might be served best if certain facts were ignored and others taken into consideration. The consequence of these differing cultural attitudes towards testimony was that cases were constantly being thrown out of court on grounds of perjury. Such conflicting views show that nineteenth European notions of truth and justice were particular discourses that became institutionalised within the system of colonial power, but were continually undermined in practice resulting in courts having little moral authority amongst the natives. Although to quote Foucault, power 'produces domains of objects and rituals of truth,'[81] these are often resisted. In fact, these particular rituals of justice had not existed in Britain for long. In eighteenth century England the law, rather than being general in its application, was tailored to each specific individual. As with the Sinhalese, questions

---

78 G.H. Hall, 'Correspondence on the subject of coffee stealing', *SP*,1873, 168.

79 R.B. Arthur, 'Correspondence on the subject of coffee stealing', *SP*, 1873, 169.

80 Rogers, *Crime*.

81 Foucault, *Discipline and Punish*, 194.

of the social position and character of the accused were seen as crucial to questions of guilt and punishment. It was the reformers' who sought to establish the nineteenth century modern standard of the uniformity and impartiality of the law.[82] Having said this, the law in Ceylon unofficially operated along racialized lines with European planters receiving different sentences for the same crimes as non-Europeans. Again, racism undermined the reformers' ideals.

Ten years later planters continued to argue that punishments handed out by the courts were inadequate, and that a public flogging in the place where the theft took place would have greater deterrent value than prison.[83] The Chairman of the Planters' Association justified the planters' position by arguing that they understood local peoples better than the government possibly could. While the government might frame laws based on a set of ideals, planters had to,

> endure the consequences of a code of laws, conceived (to use the words of a high official) in the *spirit of the times*, which being interpreted means the spirit of an advanced and highly cultured civilisation, applied to a heterogeneous people in a far lower social grade, whose habits and ideas are wholly diverse from those of the framers of that Code.[84]

He argued that the government was out of touch with local realities and racialized difference.

> [o]ur task will be to convince the local authorities, in the first place, and those in Downing Street afterwards,…that the planter here has not to contend against single-handed thieves, or with parties of 2 or 3 confederates, carrying on their work in stealth; thieves whose hand is against every man, and against whom the whole community would join. He has to deal with thieves who come in hordes, a whole village, who do not care to conceal their work except from their victims; thieves who are sure of the sympathy, even if indeed they do not enjoy the active assistance of their neighbours.[85]

While this could be dismissed as racial demonising, a reflection of what Meyer[86] terms the planters' obsession with the threat of 'native society', the following suggests that there was an anti-British ideology behind some of the theft of coffee. While much of the theft of coffee was probably seen by the perpetrators as such, this was not invariably the case. According to a police officer, 'in some villages the young men are brought up from their childhood with the idea that according to ancient Kandyan custom the highland on which the [coffee] estate has been opened belongs by right of inheritance to the village below and in helping themselves they are only taking the rent of the ground.'[87] Rogers argues that there was no cultural misunderstanding between Kandyans and British here. Rather, that 'the villagers and the British agreed in principle that 'theft' was wrong, but differing conceptions

---

82 Wiener, *Reconstructing the Criminal*, 58–61.

83 *PPA*, 1883–84, 36–37.

84 Ibid., 152.

85 Ibid., 153.

86 Meyer, 'Enclave', 213.

87 A.C. Dep, *A History of the Ceylon Police (1866–1913)* (Colombo: Police Amenities Fund, 1969), 101.

of property rights resulted in differing opinions about whether or not the taking of plantation produce was really theft.'[88] One could also argue that the villagers saw their taking of coffee as legitimate political acts of resistance to the British control over once common land.

As a consequence of their dissatisfaction with the courts, planters expressed the desire to administer justice themselves. Over and over again, planters called into question the ability of magistrates to convict thieves and called for 'the liberty of a superintendent to punish, at his discretion.'[89] They wanted permission from the Government to flog labourers on the estates. One planter wrote, 'Allow us to Lynch-law the thieves, and convict them afterwards.'[90] And in fact there were many instances of the planters simply taking law into their own hands which can be seen as acts of resistance to new forms of governmentality that were deemed far too bureaucratic and onerous.

The Colonial Secretary objected to the inference on the part of planters that they were the only commercial growers of coffee and that most coffee found for sale outside estates was stolen. There were thousands of tons of native coffee produced annually and 'it is thus shewn that there are other interests besides those of the European planter to be considered, when the question of imposing restrictions on the local dealers, who are the chief purchasers of this coffee.'[91] The Colonial Secretary dismissed the planters' requests to be granted the right to flog, confirming in effect, that the modern state could not officially relinquish its monopoly on violence.

The Framing of the Coffee Stealing Ordinances

The survey of the planters was laid before the Legislative Council in 1873 as the proposed coffee stealing ordinance was debated. Mr. Ferdinands, who represented the Burgher community in the Council, thought that rather than drafting a harsher new law, the current laws should be more effectively enforced. He opposed the introduction of what he termed an 'extreme measure … of exceptional criminal legislation.'[92] He argued that the planters should be responsible for guarding their fields and that the proposed ordinance would 'prejudice the trading interests of the vast body of the Kandyan people.'[93] He continues, the Kandyan

> has hitherto found a ready market for his coffee in his own compound, or in the nearest town, or he has taken advances and contracted to deliver his crop at a certain place within a given time, and what would be his position when his district is proclaimed in consequence of the pilfering by coolies on estates. Every bag of coffee he carries to market, or for delivery, will be suspected, examined by a prying policeman or headman and probably

---

88 Rogers, *Crime*, 69.
89 H.P. Walker, 'Correspondence on the subject of coffee stealing', *SP*, 1873, 168.
90 W. Bisset, 'Correspondence on the subject of coffee stealing', *SP*, 1873, 172.
91 A.N. Burns, Colonial Secretary, 'Correspondence on the subject of coffee stealing', *SP*, 1873, 216–17.
92 *CH*, 1873, 42.
93 Ibid., 43.

released on payment of a bribe, to be seized again further on in the journey and similarly released.[94]

His argument hit home on two grounds. First, was that the proposed ordinance was an undue interference in free trade. The planters had attempted to circumvent this issue by claiming that they were the only legitimate producers thereby criminalising the non-European population. But Mr. Ferdinands also highlighted what all concerned knew: that the government at the local level was rife with corruption. The Kandyan member, Mr. Dehigama, also opposed the bill as 'drawn up in the interests of the planters, without due regard to the interests of the natives.'[95] In spite of opposition from non-European members on Council, the Governor indicated that he would support the bill noting that he did not think that it was prejudicial against native interests. He pointed out that in framing the bill, planter input was requested but that the more extreme suggestions of the planters were not included as 'they tended to that kind of interference which has stigmatized as curtailing the liberty of the subject, crushing rising native interests, and confounding the innocent with the guilty.'[96] The Governor suggested the bill be given a two year trial[97]

The final version of what was to become Coffee Stealing Ordinance Number 8 was presented in 1874 and contained the following main points: It forbid the loading of coffee at night; loitering on coffee estates without good cause, and the purchase of coffee from labourers. It further required that purchasers keep written records of transactions, and gave increased power for magistrates to dispense with proof of guilty knowledge after a second conviction. Most controversially, the possession of green coffee without good cause became an offence, and police officers and headmen could search for stolen coffee without sworn information. The ordinance called for punishment of not more than three months in prison, or a fine of no more than 50 rupees, or no more than 20 lashes, or any two of these punishments in combination. The Governor argued that these were stringent measures and that he was unprepared to go as far as the planters had hoped. He wrote, 'it is not the province of the Government to supply coffee planters with warders any more than it is the province of the Government at home to watch over the strawberries and currents of the Chelsea market gardeners.'[98] Having said that, he clearly gave the planters most of what they wanted.

At the end of its two-year trial in 1876, the ordinance was reviewed in Council. The planting member, Mr. Downall moved that it be continued, arguing that 'it had acted as a check to coffee-stealing, without interfering with or discouraging the cultivation or sale of coffee.'[99] He went on to tell the council that having surveyed the planting districts, he could report that very little coffee was now stolen. In particular he pointed out that '[t]he itinerant Moormen, with a single bullock cart and a pair of beam scales, who were formerly always to be found about on Sundays by the roadsides, sitting under the shade of a tree, when coolies went to town for provisions were now not at all

---

94 Ibid., 43.
95 Ibid., 1873, 44.
96 Ibid., 1873, 48.
97 Ibid., 1873, 49.
98 *SP*, 1874, xiv.
99 *CH*, 1876, 121.

to be seen.' He continued, '[t]here could be no doubt these men were the means of the removal of a very large portion of the stolen coffee, and as they were frightened away, facilities to dispose of the stolen property were very considerably decreased.'[100] The Colonial Secretary added his support to these views stating that,

> [t]he Bill had a very great deterrent effect; at the same it had not added to the criminal population nor increased the numbers in our jails. It had had the effect of frightening from the coffee districts the low-country Moormen, who under pretence of trading, were known to receive and convey stolen coffee to town.[101]

At a second reading of the proposed extension later the same year, one further argument was put forward by Mr. Alwis, the Sinhalese member. The clause dealing with the possession of green coffee, he argued would 'subvert the fundamental principles of the common law and of justice.' He pointed out that 'assuming a man to be guilty until he could prove his innocence, threw the onus of proof upon the individual charged...'[102] The ordinance came up for renewal every two years, but by 1882, it was clear from Council debates that there were concerns not only about its legality, but also about its effectiveness.[103] Nevertheless it was renewed in 1882 and yet again in 1884.[104] In 1885, as coffee was increasingly being replaced by tea and cinchona, it was replaced by the broader Praedial Products' Ordinance No. 9 which again made the possession of unripe produce an offence and empowered police magistrates to sentence offenders to flogging in addition to prison.[105] The clause regarding the possession of unripe products was immediately overturned by the Supreme Court to the dismay of the Planters' Association.[106] When it was reintroduced in Council in the 1886–87 session, it was attacked by the Tamil member Mr. Rama Nathan as a tyranny. He went on to make the broader point, 'this is a most iniquitous Bill. It is all very well that the interests of the planters should be protected. Because they have taken charge of the mountain zone, is that a reason why we should deny the natives ordinary immunity from oppression and persecution?'[107]

The ordinance generated political tensions and undermined fundamental civil rights, and yet it is not at all clear that it was effective. Between 1876 and the end of 1878, records of the Supreme and Police Courts showed that 1063 people were accused of coffee stealing. Of these 625 were brought to trial, 550 under Common Law and only 75 under the Ordinance.[108] By the 1880s the number prosecuted under

---

100   Ibid., 1876, p. 121.

101   Ibid., 1876, 122.

102   Ibid., 1876, 186.

103   *CH*, 1883, 21.

104   *CH*, 1883–84, 35, 21.

105   Interestingly, Clause 3 prescribing flogging was opposed by the Attorney General but supported by two native members who argued that without this as a deterrent the ordinance would be useless. *PPA*, 1885–86, 2.

106   Ibid., xxiv.

107   *CH*, 1886–87, 73.

108   *SP*, 1879, 845.

the Ordinance increased somewhat. In the early 1880s it averaged 73 cases per year.[109] A worrying trend for the police was that there were increasing numbers of robberies of coffee stores on estates by armed gangs. The major explanation offered by the Inspector General of Police for the rise in the 1880s was the hard times brought on by the failure of native coffee and the non-payment of wages by planters on failing estates.[110] He went on to argue that Headmen and village police often overlooked and even aided robbers. The Inspector General suggested that the way to counteract this rise was through better policing. More specifically, he advocated the imposition of a police force on the 'worst villages' at their own cost;[111] the establishment of more rural police stations; and the 'bestowal of native rank' on Headmen who could better control their own people.[112] The difficulty faced by the state was that although the Headmen were expected to serve as eyes for the state in the countryside, they were unpaid or at best very poorly paid, and therefore failed to provide either detailed surveillance of the population or identification of criminals.

Conclusion

Cattle trespass and coffee stealing were rife throughout the coffee period. The entanglements of nature and power were such that the legibility of the coffee enterprise was very poor. The estates were too open, the jungles surrounding them too dense, the villages too close and the cattle too hungry and too numerous. Because of the way they had seized the highlands and occupied the common land, the British had too little legitimacy. Increasingly as coffee failed, the villagers and the labourers had too much hunger. The coffee itself was frustratingly illegible. I have described how the British tried to stem flows between the spaces of the estate and the village and how they threatened local peoples with flogging and the suspension of the civil rights. These were attempts to increase levels of surveillance, hoping against hope to render the highlands more legible. We have also seen how the planters, in this case as in the case of labour conditions and health care, continually attempted to push the government towards the framing of more authoritarian legislation, seeking continually to maintain defacto sovereign control not only over the territory of the estates, but to create a zone of disciplinary control over the surrounding villages. However, the ubiquity of cattle and the illegibility of coffee coupled with the active resistance of estate labourers and villagers alike meant that up until the end of the coffee period, planters were unable to effectively secure their borders. This only furthered feelings of abjection, of a lack of control over boundaries, and the belief that the native populations and their spaces were fundamentally corrupt and criminal by nature.

---

109 *ARC*, 1884, 51c.

110 *ARC*, 1883, 28c–29c. The collapse of coffee and subsequent impoverishment of the peasantry is the subject of Chapter 7.

111 In fact this was done temporarily as punishment in villages where there was recurrent, unsolved crime (Rogers, *Crime*, 49)

112 *ARC*, 1883, 29c.

Chapter 7

Landscapes of Despair: The Last Years of Coffee

One might say that the ancient right to take life or let live was replaced by a power to foster life or disallow it to the point of death.

Michel Foucault[1]

Introduction

In response to the demands of a greatly enlarged export economy during the nineteenth century much of the rainforest of the central highlands of Ceylon was destroyed. The British planters cleared vast tracts and cumulatively the Kandyan peasants burned even more as they sought to extend their *chena* fields for coffee production.[2] In fact, as Webb states, the coffee industry in Ceylon resulted in 'the most extensive conversion of rainforest into tropical plantation agriculture to be seen anywhere in the British Empire in the nineteenth century.'[3] In this chapter, I first outline the consequences of this large-scale deforestation and the monocultivation of coffee. I then describe the increasing governmentalization of the highland ecology as scientific solutions to ecological/economic problems were sought. The private problems of failing plantations increasingly became seen as political concerns to do with managing the collective welfare of the various populations of Ceylon.

From the late 1830s on, as more and more coffee was planted on chena land that had been periodically fallow, mixed-use peasant cropland or largely uncultivated, coffee plants became ecologically vulnerable. Monocultivation promoted outbreaks of diseases and allowed these to spread. As early as the 1840s an insect called *Lecanium* scale, known locally as 'the bug' began to spread, reducing yields to one third or less.[4] This pest was halted by the forest breaks that still existed at that time between plantations. However, by the end of the 1860s most of the forest had disappeared.[5]

In spite of periodic outbreaks of disease, in 1868 P.W. Braybrooke, the GA of the Central Province, foresaw a rosy future for coffee. Ceylon was emerging from two years of financial depression, and while drought was affecting peasant coffee in some districts, the area under coffee was expanding once again both in the peasant

---

1   Foucault, *The History of Sexuality: Vol. 1*, 138.

2   Webb, *Tropical*, 2.

3   Webb, *Tropical*, 2.

4   Vanden Driesen, 'Coffee (1)', 54.

5   Webb, *Tropical*.

and European sectors and the estates were in their best condition ever.[6] Little did anyone know that a blight of unknown origin was about to descend upon the island and within several decades destroy coffee production in the highlands and spread to coffee plantations elsewhere in the world.

Hemileia Vastatrix

The first signs of alarm came in 1869 from a planter in the south-east who sent some diseased leaves to G.H.K. Thwaites, the Director of the Royal Botanic Garden at Peradeniya (known as the Botanic Garden). Upon examining the infected plants, Thwaites found the undersides of the leaves covered in a soft powdery reddish-orange substance. These it emerged were the reproductive spores of a fungus growing in the tissues of the leaves. The spores were blown from plant to plant spreading the disease. The leaves of affected plants fell prematurely and the young fruit dried up.[7] The Director had never seen this fungus before and immediately sent specimens of diseased leaves to the Rev. M.J. Berkeley, who was considered to be the greatest specialist on fungi in Britain. Berkeley pronounced the fungus to be 'quite new to science', and named it *Hemileia vastatrix*.[8]

Subsequent research has revealed the complex interconnections between coffee and the fungus. The coffee plant originated in the Ethiopian highlands of East Africa, as did the disease *Hemileia vastatrix*. As all types of coffee show some resistance to the fungus, it is likely that they co-evolved. The coffee fungus, or rust as it is often called, thrives in warm, wet conditions. It appears that because the Ethiopian highlands have a relatively dry climate and coffee plants there were interspersed with many other plants, coffee and the fungus were able to coexist. As demand for coffee in the Muslim world began to grow beginning in the fifteenth century, coffee production expanded to Arabia, whose extremely dry climate effectively blocked the transfer of the fungus. As European consumption increased after the mid-seventeenth century, coffee plants from Arabia were transferred to Europe's tropical colonies to be grown on plantations. The Arabica coffee that diffused thus had very little resistance to the fungus. In his survey of the diffusion of coffee rust, McCook suggests that the disease spread to southern India and Ceylon in 1869 either though monsoon winds, (the spores can survive for several weeks), or more likely through human transmission.[9] One theory is that it was brought by British soldiers returning from a campaign in Ethiopia in the late 1860s or perhaps the spores could have been brought from East Africa in packing material surrounding imported ivory or spices. Alternatively, it is possible that the invention of the Wardian case[10] could

---

6   *ARC*, 1868, 34.

7   *ARC*, 1871, 531.

8   Ibid., 531.

9   McCook, S. 'The Global Rust Belt: *Hemileia Vastatrix* and the Ecological Integration of World Coffee Production Since 1850', *Journal of Global History*, 1 (2006): 177–95.

10 The Wardian case was a glass box, invented in 1835 by Dr. Nathaniel Ward. It could be sealed to contain enough water to maintain plant growth throughout a long ocean journey. This allowed live plants to be shipped long distances for the first time. Notable nineteenth

have allowed the disease to have spread from imported seedlings.[11] Once in Ceylon, the fungal pathogens were dispersed by monsoon winds and found ideal conditions – heat, high humidity and monocultivation. The uniformity of the plantations and the disappearance of the forest breaks caused the disease to become widespread and persistent rather than local and intermittent as early blights had been.[12] As we shall see, it was through a chain of unintended consequences joining Africa, Asia and Australia and eventually stretching to all coffee growing regions around the around the globe, that a disease that had been localized for centuries became a global economic force that defied all coordinated attempts to combat it.

Official Responses to the Disease

Whilst the disease was rapid in its spread, the government mobilised its resources to deal with the threat, due to the economic importance of coffee. At the nerve centre of the response was the Botanic Garden, which was in contact with botanic gardens at Kew and in other places within empire. A network of planters and AGAs fed information on the spread of the disease and responses to attempted cures to the Botanic Garden for evaluation. In his Annual Report for 1870 the GA for the Central Province noted that in parts of the low country coffee seemed to be dying out, although he added that 'on the whole, the prospects of coffee planting at the present time are good, if less bright and attractive than they were at the beginning of last year.'[13] This established what was to become a schizophrenic pattern of official response, vacillating between optimism and gloom. Although individual planters were in constant contact with the government, the Planters' Association did not officially react to the fungus at this time, focused as they were in 1870 on the issue of labour shortages and countering bad press over the accusations of Dr Van Dort about planter neglect of the labourers' health.[14] The Director of the Botanic Garden, in his Annual Report for 1871, noted that 'the rapidity with which this coffee leaf disease has spread throughout the coffee districts of the Island has been perfectly marvellous, and it is probable that not a single estate has quite escaped, though it has appeared in a very slight degree on some.'[15] In that same year, the GA for the Central Province stated that the estate crop was only a quarter of that expected and that 'much native coffee did not produce.'[16] AGAs for various districts concurred

---

century commercial successes associated with the case were the transfer of tea from China to India, and of orchids from around the world to Britain. The most notable economic catastrophe associated with it was the introduction of phylloxera from America to France and the near total loss of the European grape. On this see C. Campbell, *Philloxera* (London: Harper Collins, 2004).

11 McCook, 'The global rust belt'.

12 Webb, *Tropical*, 115

13 *ARC*, 1870, 39.

14 *PPA*, 1870–71.

15 *ARC*, 1871, 531.

16 *ARC*, 1871, 39. In fact the fungus grows best in warm, wet areas below 1400 meters, which is where most peasant coffee was found (McCook, 'The global rust belt').

that while the disease was widespread, it was most severe in the lower, warmer districts and consequently it was impacting peasant coffee gardens even more than the estates. However, the AGA of the relatively low-lying Matale district reported that the estate coffee crop had failed there as well. Nevertheless, he stated that by the end of the year the disease had almost disappeared and that the prospects for the following year were bright.[17] Such was the nature of the disease, which ebbed and flowed, that hope was held out over and over again that the worst had finally passed. In its 1871–72 Annual Report, the Planters' Association noted its concerns over the impact of leaf disease and expressed the hope that the problem could be solved by the Botanic Garden.[18] But again, their attention was directed towards what seemed to them to be more pressing political issues regarding the Medical Ordinance discussed in Chapter 5. The following year there were reports that the prospects for estate coffee had improved in large part because of high prices in London. Although the average yield had dropped dramatically from a high of 5.9 to 3.21 cwt per acre, the area under coffee was expanding (see Table 2.1). However, peasant coffee again did very poorly and in the district of Uva it was 'almost a failure.'[19] The great reduction in peasant coffee was beginning to have an impact on the well-being of villagers.

The Director of the Botanic Garden in 1872, indicated that the disease was everywhere and that 'although there is probably little hope of the disease ever quite disappearing from the Island, yet it seems to be pretty generally believed that under good cultivation the loss occasioned by it is not nearly so serious as was at first feared might be the case.'[20] The Director's mistaken optimism is understandable given the complexity of the disease. In the first place, the disease was not uniformly devastating. It affected some areas more than others. Furthermore, it would appear to go away, only to return at a later date. Second, it was less virulent in young, healthy trees and at higher altitudes with a pronounced dry season. Third, it rarely killed the coffee trees; it weakened them and reduced their yields. Finally, the disease could be partially controlled through manuring and applying shade.[21] Many of these factors, in the early years at least, created the illusion that the disease could somehow be contained. In spite of the fact that the disease was never effectively controlled, coffee production expanded in 1873 with new land selling at unprecedented prices due to the availability of capital and the high prices of coffee on the London exchange.[22] And yet in spite of the great increases in acreage of coffee after 1870, because of drastically dropping yields, production never again reached the 1870 level.

By 1873, the planters were expressing grave concern about the impact of the disease on their crops. But it was clear to them that the government was doing all it could to solve the problem.[23] In addition to searching for a cure, the Director of the

---

17 *ARC*, 1871, 52–55.

18 Ibid., 82.

19 *ARC*, 1872, 52.

20 Ibid., 434.

21 Clarence-Smith, 'The coffee crisis', 103.

22 Forest land that might have sold for two pounds per acre ten or 20 years before was selling in the 1870s for up to 28 pounds per acre. Vanden Driesen, 'Coffee (2)', 164.

23 *PPA*, 1873–74, x. The Planters' Association was set up as a lobby group in 1854, in order to channel complaints about the lack of expenditure on roads during the administration

Botanic Garden reported that J.D. Hooker of Kew had arranged for West African coffee plants (*coffea liberica*) to be sent over to ascertain if they were disease resistant.[24] In 1874, the AGA of Badulla noted that although the crop was light and disease was widespread, the high prices of coffee had created a boom in the opening up of new coffee lands by both Europeans and non-Europeans. One side effect of this was a deluge of court cases brought by the government against villagers for encroachment on Crown Lands.[25] By that year, it had become clear that the imported West African coffee plants were also being attacked by the fungus, although growers persisted in planting them for the next decade.[26] And yet in spite of this setback, the Governor in his address to the Legislative Assembly found cause for optimism in the fact that investors were continuing to convert forest to coffee. He said:

> it is gratifying to see how much confidence is felt in the permanency of a successful and profitable coffee cultivation in the Island, as is sufficiently shewn in the competition for land suitable for planting. It cannot be denied that the leaf disease, which still prevails in different districts, does cause a corresponding diminution of crop, and that were it not for the presence of this disease our prospects, with the high prices now obtained for coffee, would be still more encouraging.[27]

He added that in order to learn as much as possible about the disease, the Colonial Office had ordered that a questionnaire be sent to all colonies where coffee is raised. While hopes for coffee had clearly not been abandoned, the Botanic Garden encouraged the spread of tea at higher elevations and 690,000 cinchona seedlings as well were distributed during that year alone.[28] The following year the Botanic Garden distributed thousands of tea plants and continued its tests on Liberian coffee, concluding that, although attacked, it suffered slightly less from fungus than Arabica.[29] The Chairman of the Planters' Association told the members in his annual address in 1875 that the disease was 'still ineradicable', but pointed out that old and weak coffee was most severely attacked.[30] The implications of this were that through proper management of estates, the impact of disease could be somewhat mitigated. This rather hopeful note was echoed in the annual report of the GA for the Central Province in the following year. He wrote that 'owing to the better system of cultivation in force at present, planters are able to make a successful stand against leaf disease.'[31] In light of the inability of botanists to determine a remedy for the disease, planters experimented with applying very high doses of manure, believing that this would strengthen the bushes. The director of the Botanic Garden cautioned

---

of Governor Anderson. Vanden Driesen, 'Coffee (2)', 157.

  24 *ARC*, 1873, 51.
  25 *ARC*, 1874, 42.
  26 Ibid., pt iv, 1.
  27 *SP*, 1874, xiii.
  28 Ibid., xiii.
  29 *ARC*, 1875, 39.
  30 *PPA*, 1875–76, 2.
  31 *ARC*, 1876, 7.

them against this course of action, but had no alternative to offer.[32] With continuing high prices for coffee in the London market both Europeans and peasants invested in new coffee land. Increasingly, Liberian coffee was being planted in the lower districts, where it was thought to do better than Arabica. There were also reports from some districts that the disease was abating and the crop was expected to be good in the following season.[33] In that same year the Botanic Garden set up a new experimental garden in the low country to raise Liberian coffee for distribution.[34] By 1878, while coffee above 2,000 feet was still producing reasonably in some cases, that below produced very little. The AGA of Matale reported that peasant gardens produced the smallest crop ever.[35] The government began distributing Liberian coffee plants to peasants free of charge and the Botanic Garden imported West Indian coffee seedlings in the hope that they might be resistant.[36]

To compound misfortune, however, during the late 1870s as disease ravaged production in the highlands, a depression struck world economies and was especially devastating to colonial export economies. Consequently, demand for coffee dropped greatly at a time when planters were struggling with reduced yields. During the last years of the 1870s there was increasing distress amongst the peasantry as their coffee failed earlier and more completely than much of the estate coffee at higher altitudes.[37] By 1880 the experiment with West Indian coffee had also failed and most of the 26,000 Blue Mountain seedlings imported from Jamaica had been killed by the disease. The Liberian coffee was also proving even more susceptible to the fungus than had been thought. The government reported that although coffee was still the staple, that the disease was everywhere and the crop increasingly small. As coffee yields declined, estates began to fail. These abandoned (shuck) estates then became key transmission sites from which the disease could spread even more freely.

The cultivation of tea and cinchona, on the other hand was increasing, but not sufficiently to make up for the heavy losses in coffee.[38] By the early 1880s the tone of official reports on coffee were darkening. Marshall Ward, a specialist in fungal diseases who had been sent by Kew to Ceylon to study the leaf disease, concluded that there was little prospect that a cure could be found.[39] In his report he issued the following warning against monocultivation, '[h]aving provided immense quantities of suitable food [in the form of the coffee plant], carefully preserved and protected, man unconsciously offered just such conditions for the increase of this fungus as favour multiplication of any organism whatever.'[40] Ironically, the Botanic Garden, which was used for research into diseases and their prevention, unintentionally

---

32 *ARC*, 1877, 1c.
33 *ARC*, 1876, 17, 183.
34 *CH*, 1876–77, 239.
35 *ARC*, 1878, 45.
36 Ibid., 2c.
37 *SP*, 1880, v.
38 *ARC*, 1880, 39–40, 286c.
39 *PPA*, 1881–82, xxvii.
40 M. Ward, 'Research on the Life History of *Hemileia Vastatrix*, the Fungus of the "Coffee-Leaf" Disease.' *Journal of the Linnean Society of London – Botany* 19 (1882), 334–35. Quoted in McCook, 'The global rust belt'.

fuelled the ecological disaster. For some of the Liberian coffee seedlings that were imported in Wardian cases by the Botanic Garden contained the Lecanium scale bug, which spread widely to Ceylonese coffee and in the mid-1880s delivered yet another blow to the industry. And if all of this wasn't enough, coffee was also greatly affected in the 1880s by another imported pest, called the 'grub', which the Botanic Garden identified as a relative of *Melolontha vulgaris*, the common European cockchafer.[41] Further compounding the problems of these pests, disease and decreased demand was a surge in production of low-grade Brazilian coffee in the 1880s, which had the effect of decreasing world prices. This coincided with the 'Great Depression' of 1879–84 in Britain and a fall in the standard of living that prompted British consumers to switch to lower-grade Brazilian coffee at half the cost of Ceylon coffee.[42] This was the final in a series of ecological and economic blows to Ceylon coffee knocking it out of the international network of coffee producers. Ceylon had fallen to this state from being the third largest exporter of coffee in the world in 1870. In the next three sections of this chapter I will outline the impacts of the failure of coffee on the planters, the estate labourers and the peasants.

The Collapse of the Coffee Estates

By the early 1880s there was little optimism that a recovery was possible. The AGA of Badulla summed up the mood when he wrote that 'it appears evident that the district must shortly look to new products almost entirely for profits.'[43] This view was shared by the Director of the Royal Botanic Garden at Kew, who felt that Ceylon coffee would never recover.[44] Kew estimated that the total financial loss due to the disease between 1869 and 1878 was 12 to 15 million pounds sterling. Planter opinion was divided about what the future held. For many the end of the coffee industry was the end of planting for them. Others such as J.L. Shand, the Planting Representative on the Legislative Council, thought that there would be a bright future for planting in tea, Liberian coffee, cinchona, and cacao. The Planters' Association, remembering that planter speculation and mismanagement in the 1840s helped precipitate the collapse of coffee at that time, were keen to exonerate planters by arguing that the failure this time was brought on by 'a visitation which no human wisdom could have foreseen.' To hammer the point home, the planting representative wrote that if the planters were at fault, it was that they had been 'too hopeful, too trustful – faults not unnatural in hardworking, healthy men …'.[45] All of this was a thinly veiled appeal for government financial assistance. The governor expressed his regret at the decline of the planters, but argued that:

> legislation dictated by a desire to furnish direct assistance to industrial enterprises is always difficult and too often actually mischievous; but it is almost impossible to repress

41 *ARC*, 1884, 9D.

42 Vanden Driesen, 'Coffee (2)', 166.

43 *ARC*, 1881, 62A.

44 *SP*, 1881, 27.

45 *PPA*, 1882–83, 106–07.

a wish to discover some means not inconsistent with sound economical laws by which encouragement might be given either to the prosecution of new and promising industries, or to the efforts made to check the ravages of leaf disease.[46]

In August 1883 the Chairman of the Planter's Association, using Darwinian language read the obituary for coffee at the General Meeting of the Association. He said,[47]

> The depression which has so long existed with such dreadful effect on our enterprise still continues. The abnormal seasons and pests which have followed in their wake, and all those invisible enemies which have presented themselves to our notice, have baffled all our efforts to counteract them. They are no doubt obedient to laws of which we can form at present no conception. We find that not only whole races of plants, but of animals, and even the human race, have disappeared from the face of the earth in the course of past ages, owing to the altered conditions of the times; and therefore, although the laws are inscrutable, they are no doubt benign; and if we were to, as now appears likely, to suffer the almost entire extinction of the staple upon which we have so long depended, it will not be a new thing in the history of the world, but will only be further proof of the rule of laws which are to us inscrutable.

The first half of the 1880s was the worst for the planters. By mid-decade many planters followed government advice and made the transition to other crops. In 1886 the GA of the Central Province wrote,

> [t]he rapid conversion of abandoned coffee estates into tea gardens is a notable fact in every part of the hill country. Vast expanses of tea now meet the eye in every direction...Factories [for tea curing] are being rapidly erected everywhere, giving remunerative employment to large numbers of sawyers, carpenters, and masons. New bazaars are springing up along the lines of roads, and there is every appearance of returning prosperity.[48]

To which a superintendent of a coffee estate in Matale Districts added, '[t]here is now little coffee left in this district, ... As a rule, coffee is merely tolerated where it does not interfere with other products, and cannot be said to be cultivated; it receives no pruning worthy of the name, and suckers are often allowed to grow.'[49] In the Annual Report of the Planters' Association for 1886, the Chairman called the prospect for planting bright, especially for tea, 'which has practically become the distinctive staple of Ceylon.' But he went on to warn that 'it is to be hoped that the policy of not depending upon one product will be prominently kept in view.'[50] Two years later, the Chairman said before the Annual Meeting,

> it is now three years since you did me the honour of placing me in the chair, and those three years have been eventful years in the history of Ceylon, in so far that the staple product of the country has ceased to be coffee and has become tea, and that the planting

---

46 *SP*, 1883–84, xxi.
47 *PPA*, 1883–84, 33.
48 *ARC*, 1886, 22A.
49 Ibid., 23A.
50 *PPA*, 1886–87, xxviii.

enterprise which then stood discredited and mistrusted has, in that brief space of time, risen again, and is rapidly approaching its former prosperity.[51]

But not all coffee planters were able to make the transition to other crops. In the late 1880s an old coffee hand had this to say as he saw the industry collapse around him:

> [m]any of the plantations were deserted, the capitalists took fright, and superintendents were thrown out of employment and set off to other countries. There was a regular migration to Northern Australia, Fiji, Borneo, the Straits, California, Florida, Burma and elsewhere. I should say that out of 1700 planters we lost at least 400 this way.[52]

At times, planters were so bankrupted that they need charity from groups such as the Kandy Friends-in-Need Society to buy their ticket home.[53] But although most coffee died out on the island, the disease survived and moved on. Following the networks of planters and labourers, the spores spread to southern India and eventually to South Africa, East Africa, Java, Fiji, and Australia.[54] The coffee rust struck the Philippines, the world's fourth largest exporter of coffee, in 1889 and by 1892 the island no longer produced sufficient coffee for export.[55]

The Impact of Coffee's Failure on Plantation Labour

As plantations began to fail, wages were not paid. While non-payment had been a sporadic problem throughout the coffee period, it was aggravated by the coffee disease and the changing fortunes of the planters. The Police Report for 1882 stated that there had been 'frequent complaints of the loss of poultry and pigs in the planting districts, but the belief is that they are mostly stolen by estate coolies for food, the coolies' pay being generally several months in arrear.'[56] Increasingly labourers deserted, in spite of the fact that Labour Ordinance No. 11 of 1865 punished desertion with a term of up to three months of hard labour. But not all labourers who were not paid deserted. The Lieutenant Governor tabled a dispatch before the Legislative Council in 1883 which described the dilemma that estate labourers on failing estates faced in the early 1880s. After 1879, as the price of coffee dropped in the international market and yields continued to fall, labourers' wages fell seriously into arrears and planters often only had three or four days work per week for their labourers. As three days work a week would only cover the worker's rice ration, most ended up simply working for food. Workers were trapped on the estates, because if they left they had virtually no chance of recovering their back wages. Furthermore, they had very limited alternatives as the general depression meant it was highly unlikely that they could get another job and there was hunger back home in southern India. So they

51 *PPA*, 1888–89, 2.
52 J. Ferguson, *Ceylon in the Jubilee Year* (London: J. Haddon and Company, 1887).
53 *PPA*, 1881–82, xxix.
54 Webb, *Tropical*, 116.
55 McCook, 'The global rust belt'.
56 *ARC*, 1882, 23c.

stayed and hoped for better times.[57] The government became so concerned about the exploitation of the labourers that they sought to amend Labour Ordinance No. 11 of 1865. The Colonial Secretary's Office wrote to the Planters' Association in October 1883 informing it that in light of frequent reports of non-payment of wages, the government felt the necessity to protect the interests of what he termed 'an ignorant class of labourers.'[58] The government decided on this course of action based in part on a number of unsuccessful suits for unpaid wages brought by labourers and kanganies against estates during 1883. The Secretary of the Planters' Association replied to the Colonial secretary, that while it might appear that wages were not being paid, that in fact this was a misconception. He claimed that the planters were simply holding the labourers' savings for them! While admitting that there were some cases of arrears in wages, these were represented as exceptions to the rule. The Secretary wrote, 'it is needless to remind the government of the desperate struggle the planters as a body have had during the last few years, owing to the terrible and persistent failure of their crops...'.[59] He concluded by stating that the Association felt 'deeply aggrieved that representations so prejudicial to the enterprise and character of the planters', had been made to the Home government and argued that any new legislation was unnecessary.[60] In December 1883 the Colonial Secretary's Office sent the Planters' Association a letter arguing that the proposed legislation had three principal goals: to guard labourers against a loss of wages due them, to provide a means for labourers to recover lost wages, and to provide the government with a way of knowing if wages were in arrears.[61] The planters particularly objected to the third part of the proposed ordinance requiring them to report to the government at the end of each month the number of labourers in their employ and whether any wages were in arrears. This was decried as undue governmental interference in the relations between employer and employee. When the Ordinance, known as the Cooly Wages Bill came before Council, J.L. Shand, the member for the planting community, rejected those clauses concerning the requirement that planters report if wages were paid, and those which imposed sanctions against planters who failed to report wages paid. The Governor backed down in the face of planter opposition, saying that while he believed that the clause requiring the planters to inform the government monthly about the state of wages was necessary to the working of the Bill, that he was prepared to remove those clauses which specified penalties for non-compliance. Rather, he said, he would trust the planters to follow the new law. He argued this on the grounds that

> there can be no doubt that a law worked with the cordial good will of those whom it concerns, is far more efficacious than it can be made by the most rigid system of inspection or under the guard of the severest penalties. Inspection and acts of minute interference are in themselves evils. Sometimes, perhaps even often, they are necessary evils – but still evils – tending to excite irritation, to promote opposition and to produce

---

57 *CH*, 1883–84, 21.
58 *PPA*, 1883–84, 161.
59 Ibid., 167.
60 Ibid., 168.
61 Ibid., 177.

actual indifference to abuses, if not almost to endear them, by rendering them objects of conflict and defence.[62]

In spite of the teeth being removed from the Ordinance, the Chairman of the Planters' Association rejected it as 'altogether out of proportion to the circumstances of the case, and reflect[ing] unwarrantably on the character and honesty of the Planters as a body.' He continued, 'if passed into law it would disturb the happy relations hitherto subsisting between our labourers and ourselves, which have been most beneficial to the country, and advantageous, above all, to the coolies'.[63] He went on to suggest that the government withdraw the bill.

In order to scuttle the ordinance, the Chamber of Commerce and the Planters' Association attempted to get the legislation broadened to include all labourers on coconut plantations and in the construction trades. Such a broadening would have removed the implication of wrong doing from the coffee planters and would have so increased the scope of inspections that it would have no doubt swamped the small bureaucracy assigned to it. The Colonial Secretary opposed the measure on the grounds that legislation was only necessary on the coffee estates; because that was where abuses had taken place and also because of the special care that the government had to take of the immigrant labourers.[64] The Planting Member challenged the necessity of the legislation arguing that 'at no previous period during which the immigrant labourer has been employed in Ceylon has the condition of the cooly been in such a favourable state as it is at present.'[65] He referred to the proposed government checks on whether labourers had been paid as a 'system of espionage between employer and employed.'[66] In February 1884 the bill (Ordinance no. 16) was passed without opposition. While the planters had to disclose the state of their arrears quarterly and labourers were more easily able to sue for non-payment of wages, the bill had been effectively gutted of its force. After its passage the outgoing Chairman of the Planters' Association told the Association that the planters' opposition to the original form of the bill had modified it for the better and that the planters' opposition to the government's action could be described in the following way:

> there is always a practical and a theoretical view in almost any subject connected with legislation ... and it is for the government to take a theoretical view and it is for them to

---

62 *CH*, 1883–84, 14–15.

63 *PPA*, 1883–84, 179.

64 *CH*, 1883–84, 39–40. While it is true that the Government of India took an interest in the welfare of the labourers, they were also very much influenced by notions of laissez-faire and had an eye to Ceylon as a site to siphon off excess labour. Consequently, in 1880 the Madras Government turned down a suggestion from the Ceylon Government that the licensing of kanganies would cut down on abuses. The Chief Secretary of Madras wrote, 'With regard to the general question of organizing recruiting for the Ceylon labour market, this Government is disposed to doubt whether free action, such as has always prevailed hitherto, is not the most advantageous to the island and to the labourer.' (Chief Sec. of Madras to Col. Sec. of Ceylon, No. 831 of 1 June 1880, Proceedings. No. 6 of June 1880, Madras Public Proceedings, *Vol.* 1555, p. 239, quoted in Wesumperuma, *Indian Immigrant*, 25.

65 *CH*, 1883–84, 41.

66 Ibid., 42.

submit their theory, when laid down, to the practical test of those who, like yourselves, know thoroughly well – far better than other people can – the nature of your own business.[67]

Once again, the planters demonstrated that they were able to resist what they saw as illegitimate governmentality. They reaffirmed their defacto sovereign power within the space of the estates.

The Failure of Peasant Coffee

While the planting community as a whole, if not all individual planters, could make the painful transition from coffee to tea, the move was much more difficult for peasant smallholders. For the peasantry the failure of coffee had been catastrophic. Not only did coffee collapse earlier for the peasantry, but their distress was compounded by changes in taxation policy. The Paddy or Grain Tax was set at one tenth of the rice crop and was commuted in 1878 to a cash payment, with disastrous consequences. The Paddy Tax and the Road Ordinance[68] were taxes of the sort recommended by the Colebrooke-Cameron reforms in 1833 to replace *rajakariya*, the system of labour owed to the king during Kandyan times.[69] These taxes were part of a package of reforms whose goal was to modernise the state and force the peasantry to work on the plantations. Instead, the peasantry turned to coffee as a cash crop. Although initially surprised by this outcome, the government encouraged peasants to grow coffee, and peasant production doubled between 1820 and 1824, and doubled again between 1826 and 1833. In fact, peasants had produced the majority of the coffee grown in Ceylon into the late 1840s. As peasants produced more and more coffee, the government, aware that peasants had more cash, set the Paddy Tax unrealistically high.[70] The upshot was that peasants could only pay their taxes if the rice harvest was very good or if coffee was grown as an important supplementary cash crop.

---

67 *PPA*, 1884–85, 3.

68 Under the Road Ordinance 8 of 1848, every male aged 18 to 55 had to work six days per year on the roads or pay an annual commutation of 1.5 rupees. Labourers living on estates and monks were exempted from this tax. The peasants hated this tax and often chose to commute it even if they did not have the money to pay the commutation. So common was non-payment that the jails were full of road ordinance defaulters who preferred prison to a week's work on the roads. Nevertheless, the funds raised by the commutation were significant, in 1867 constituting 15 per cent of the total funds for public works. (Bandarage, *Colonialism*, 249–50).

69 In 1832 there were 3,000 labourers working in the Kandyan provinces on public works. (de Silva, *Ceylon Under the British*, 441). The Road Tax continued the practice of having the peasantry pay for road building, but used a more modern means to effect that payment. The peasantry considered it to be a revival of *rajakariya*. The people of Dumbara signed a petition to the Governor against this tax. In it they said, 'We cannot understand what crime we have committed to deserve so great a punishment.' (Petition of 6 July 1848 from the people of Dumbara enclosed in Torrington to Grey of 9 July 1848, quoted in De Silva, *Letters on Ceylon*, 10).

70 Bandarage, *Colonialism*, 143.

Through the 1860s, the peasants were able to produce sufficient coffee to pay their paddy taxes.

Table 7.1   The decline of peasant coffee, 1870–86[71]

Year	Exports (cwts.)
1870	132,523
1871	170,397
1872	140,623
1873	121,577
1874	96,758
1875	114,431
1876	79,843
1877	82,282
1878	46,237
1879	54,414
1880	44,753
1881	29,769
1882	35,499
1883	14,823
1884	11,886
1885	20,611
1886	10,046

In the early 1870s however, a situation was beginning to develop that was to become increasingly grave for the peasantry over next several decades. The AGA of Nuwara Eliya in 1872 reported that the amount of paddy commutation tax collected was much less than the previous year due to the poverty of the peasantry in many parts of the district.[72] This was the beginning of a downward cycle where struggling peasants with decreasing coffee yields put off taxes in the hope that they would be able to pay them when coffee recovered. However, peasant coffee never did recover.

As the leaf disease worsened in the late 1870s and early 1880s the great depression set in. The plight of the peasantry was greatly compounded by a government decision at the time to attempt to recoup some of the losses in government revenue due to the failure of coffee by more rigorously collecting peasant taxes. AGAs were now told to collect back taxes as well as current taxes. Ordinance No. 5 of 1866 had been drafted precisely to deal with such a situation. It allowed for the expropriation and sale by auction of a peasant's land if his taxes fell in arrears. Such a policy was to have disastrous consequences for the peasantry of certain areas. When discussion of collecting arrears on the Paddy Tax came before the Legislative Council in 1878, non-European members raised the issue of the regressive nature of taxing food, pointing out that there was nothing comparable in England. The Queen's Advocate argued in reply that while it would indeed be wrong to impose a grain tax in England because

71 Vanden Driesen, 'Coffee (2)', 164, 168.
72 *ARC*, 1872, 68.

property was unevenly distributed, that in Ceylon property was evenly distributed and conditions of life were equal and simple.[73]

While it was true that the highlands of Ceylon were characterised throughout the nineteenth century by a more equitable distribution of land ownership than in Britain, such arguments dovetailed neatly with the 'myth of the lazy native' and environmental determinist views of the dire moral consequences of tropical abundance. For example, Steuart in 1855 contrasted Ceylon with England in the following terms:

> In England the study of statesmen is to find employment for the poor; while in Ceylon the difficulty is to find poor to employ. England has not sufficient land to produce food for its manufacturing people; while Ceylon has not sufficient labouring population to cultivate the soil for English capitalists and has none to spare for manufacturing purposes.[74]

To which Governor Robinson at the opening of the Legislative Council in 1866 added:

> [t]he wants of the native population of the island are few and easily supplied by an occasional day's work in their own gardens or paddy fields. Their philosophy, their love of ease and indolence or their limited ideas, whichever may be the real cause, render them perfectly contented with what they already possess;...[75]

One can see how such views could lead conservative government officials to argue that the hunger of the peasants was simply due to their lack of willingness to work, and that to reward this by not collecting taxes was to reward laziness.

The new GA of the Central Province, John Dickson was one of those who blamed the peasants for their own poverty, and in 1882 ruthlessly began to collect arrears on the Paddy Tax. The results were evictions and starvation in some districts.[76] But whereas Dickson was the driving force behind the implementation, there were plenty of non-Europeans who were keen to profit from the misfortune of their fellows. Bandarage argues that Headmen, as assessors of the value of fields for the British

---

73 *SP*, 1878, iv, cited in Bandarage, *Colonialism*, 146. In this, the Queen's Advocate echoes the Secretary of State, Earl Grey in 1848, who though opposed to food taxes in Britain, argued that the East was fundamentally different from Britain. He 'denied that direct taxation was injurious in its effect in Eastern countries: the masses needed and consumed so little that it was the only method of taxing them;' such taxation was necessary if they were to support 'those institutions and that machinery of government...[so] essential to progress, and even to the maintenance of civilized society'; direct taxation was 'conducive to [the] true welfare' of the native peoples in that it served as means of prodding them to exertion.' (CO 55/91, Grey-Torrington, No. 305, 24 October 1848), quoted in M. Roberts, 'Grain Taxes in British Ceylon, 1832–1878: Theories, Prejudices and controversies', *Modern Ceylon Studies* 1 (1970), 122.

74 J. Steuart, 'Notes on Ceylon and its Affairs During a Period of Thirty Eight Years Ending in 1855', quoted in R. Pieris, 'Society and Ideology in Ceylon During a "Time of troubles," 1796–1850, Part 3', *University of Ceylon Review* 10 (1952), 81.

75 Cited in Bandarage, *Colonialism*, 175.

76 Wesumperuma, D. 'The Evictions Under the Paddy Tax and Their Impact on the Peasantry of Walapane, 1882–1885', *The Ceylon Journal of Historical and Social Studies* 10 (1970): 131–48.

,were 'placed in an ideal position to manoeuvre land sales to their advantage.' He added that, [t]here is plenty of evidence which shows that they did so, and that a great proportion of the lands of defaulting peasant cultivators were bought by headmen and their relatives at very low prices.' He points out that the second most common purchasers were low-country Sinhalese and Moors.[77]

In 1883 the AGA of Badulla, the district hardest hit by expropriations, wrote that in 1879 villagers suffered a sharp economic decline due to the total failure of coffee for the previous four years.[78] By 1885 the AGA wrote that:

> by far the greater number of the native [coffee] gardens have died out completely. For the most part, what was once flourishing gardens is now a wilderness of dry sticks. It is no uncommon thing to see coffee gardens being felled for the cultivation of kurakkan...The present state of things contrasts painfully with the luxuriance of native coffee in former days, when it put in circulation among the villagers in this District as much as Rs. 600,000. It was this that ensured the regular and cheerful payment of taxes...[79]

During the 1880s there was a rise in the incidents of crime. In 1882 the AGA of Nuwara Eliya, G.A. Baumgartner wrote that Sinhalese villagers were breaking into the stores, stripping coffee trees on estates, and robbing labourers carrying head loads. Such practices were equally prevalent in the villages themselves where 'growing crops and produce were stolen to such an extent that complaints became constant.' He concluded, 'undoubtedly the pressure of hard times has driven many persons into criminal courses.'[80] The Inspector-General of Police in his annual report on crime in 1883 agreed, attributing the increase in coffee theft to the failure of peasant coffee and the unpaid wages of plantation labourers.[81] It was hunger which lay behind this spike in crime. But this view was contested by J.L. Shand, the Planter representative on the Legislative Council. He said, in arguing for the renewal of the Coffee Stealing Ordinance in 1883, that:

> the nature of these agricultural thefts throughout Ceylon is a particular one; they are not committed as agricultural thefts throughout the world generally are by some starvation-stricken creature who rushes to gratify the cravings of appetite into a field of corn or turnips, but it is a system of robbery by which many men earn a systematic livelihood. These robberies were not committed by single individuals, but by whole communities...and there were reasons to fear that they were connived at by the headmen of communities.[82]

The Planter's Association determined that the cause of the increase of theft in the early 1880s was the 'insufficiency of the punishment inflicted in cases of conviction and the weakness of the police.' Consequently, it urged the government to allow 'more summary and effectual measures', specifically flogging as 'the only effective

---

77 Bandarage, *Colonialism*, 148.
78 *ARC*, 1883, 27a.
79 *ARC*, 1885, 61A.
80 *ARC*, 1882, 72A.
81 *ARC*, 1883, 29c.
82 *CH*, 1883–84, 17.

and deterrent punishment for such crimes.'[83] And yet as the planting member spoke, the following report of famine came in from the planting districts. The AGA for Badulla wrote to the GA of the Central Province in July, 1883 that:

> I have in the last nine days, while walking through the villages of Gampaha and Medapalata for the purpose of seizing property for arrears of Paddy Tax, seen much of the state of things in this respect in these two Korales. I have visited a great number of the distressed families in their homes and found them all in similar condition – without any food except the leaves, fruits or roots of wild plants. Many persons were utterly prostrate from want of food or from disease supervening the want of food – the very young children especially. All were greatly emaciated. I learned that several infants had died in Tuppitiya through their mothers being unable to afford them natural nourishment, and I saw several children in an advanced stage of decline from this circumstance. Arrangements were made by me for supplying them with cow's milk. …Upon my arrival here, I started a relief kitchen… The number of applicants has gradually increased from 16 to 150, and I am told, as I write, that applicants are still flocking in, chiefly women (widows as a rule) and children. All the men, except two or three, whom I have required to work for their food, are sick and incapable of work – most of them very old men. All these persons are from villages within a two or three mile radius of this place. The greater number of them say that they have left members of their families at home too weak to be removed.[84]

The AGA went on to say that he was told by the Ratemahatmaya that the same conditions prevailed in other districts. It is clear that the AGA felt sympathy for the starving peasants, nevertheless, he concluded his report by blaming them for their destitution:

> It is probable that few sales will result in much change for the worse as regards the owners, [of expropriated land] while the good of the greater number will, there is [reason] to anticipate be advanced. A section of the community composed of pauper proprietors who are unable to cultivate their fields for want of energy and want of means and who instead of benefiting their neighbours are a positive burden to them, will be got rid of. If they behave like sensible people, they will thus become of much more use to themselves and to the community in their changed circumstances. There is plenty of work for them, if they choose to seek it.[85]

The planting press shared this view, arguing that 'the pinching of the stomach is morally good because it will induce the peasants to work on plantations.'[86] The above reveal the resilience of the stereotype of the lazy native, for knowing that the policy of expropriation made these people destitute, and that the collapsed coffee industry made it very unlikely that they could find work, they nevertheless asserted that the fault for their destitution was largely their own. The image of the tropical environment as conducive to ease of living hung over these debates. Perhaps the

---

83 *PPA*, 1881–82, 211, 213.

84 *ARC*, 1883, 27a.

85 Ibid., 26a.

86 A.M. and J. Ferguson, *Taxation in Ceylon with Special Reference to the Grain Taxes: The Important Duty on Rice Balanced by a Local Levy and Proposal to Substitute a General Land Tax*, cited in Bandarage, *Colonialism*, 180.

most extreme instance of such a view was put forward by the Inspector General of Police who saw even death in such an environment as pleasant. At the height of the famine, he argued that death in Ceylon was 'not so dreadful as it sounds in England where the atmosphere of an ordinary bedroom is less genial than the open air throughout the greater portion of Ceylon.' On the island, he continued, they 'slip painlessly out of life in the place where they have lain down to rest or sleep.'[87]

Three years later, James Irving, AGA in Badulla wrote that the situation of peasants in Uva was 'very similar to that of the Irish peasantry after the potato famine – except being sold up by the government for taxes instead of by big proprietors for rent.'[88] The most outspoken opponent of the Paddy Tax and the subsequent expropriation of peasant lands was C.R. LeMesurier, the AGA of Nuwara Eliya district. In his administration report for 1887 he wrote of a conversation that he had along the Badulla Road with a destitute family:[89]

Q. Who are you?
A. We are the people of Walapane.
Q. Why have you left your village?
A. Because we have lost all our property.
Q. How did that happen?
A. We are allowed to run into arrears with our Paddy Tax for two or three years. It was called up all of a sudden. No mercy was shown us, and all our property was sold.
Q. What are you doing now?
A. We are in search of employment and means of livelihood.
Q. Where are you going to?
A. Nowhere in particular.
Q. Why do you not go and work on a tea or coffee estate?
A. We have tried this, but we never got anything for our work, all our earnings were taken by our Kangani for debts he said. It is "debt" always "debt" with us.

During famines such as that of the late 1870s and 1880s governmentality was increased. It became government policy to offer food relief to women and children, the old and the sick. Able-bodied men were only offered food if they worked on government projects such as road or irrigation works. Such was the Kandyan loathing of this sort of work that AGAs reported villagers living on a starvation diet of jungle fruit and leaves rather than do what they considered degrading work.[90] AGAs also reported that it was common for able-bodied men to leave women, children and the aged in villages and depart for other districts. This was done knowing that the government would offer relief to those left behind. Aelian King, the AGA of Badulla, was so incensed by what he saw as this strategy of work avoidance that in 1879 and again in 1884 he advocated special legislation whereby able bodied men

---

87 *ARC*, 1880, 32b.

88 Irving quoted in D.M. Forrest, *A Hundred Years of Ceylon Tea, 1867–1967* (London: Chatto and Windus, 1967), 107. For an interesting account of the Great Irish Famine and governmentality see, Nally, "'Eternity's commissioner'", 313–335.

89 *ARC*, 1887.

90 On a similar unwillingness of Irish men to do such labour during the Great Irish Famine, see Nally, "'Eternity's commissioner'", 325.

who refused to work on government projects were either to be treated as prisoners and forced to work or treated as vagrants and prevented from leaving their villages in time of famine, unless they could prove they had work in another district. Such draconian measures were rejected by the government as unworkable.[91]

Throughout the 1880s, the government continued to collect of taxes on food, even if it resulted in destitution and the expropriation of peasant lands. But in 1889 this began to change with the publication in the *Manchester Guardian* of a letter by Mr. C.S. Salmon of the Cobden Club. In this letter, which had shortly before appeared in the *Ceylon Mail*, the author charged that 2,889 paddy fields had been sold by the government between 1882 and 1885 for non-payment of the Paddy Tax and that subsequently, 1048 of the villagers evicted had died of starvation.[92] John Dickson, the former GA of the Central Province sought to refute this charge in a letter to the Colonial Secretary on 25 September 1889, arguing that while, under his leadership, mistakes had been made in harshly enforcing the expropriation of property, that few of the deaths could conclusively be traced to peasants' removal from the land. The GA argued that the inability to pay taxes stemmed from the failure of peasant coffee, a natural calamity that could not be blamed upon the government.[93] His AGA, LeMesurier, radically qualified this assessment of the distress in Walapane arguing that:

> [t]he causes of the large arrears outstanding at the beginning of 1881 are now clear. They were first, the over-assessment of the fields; second the injury to the water supply by the clearing of forests for estates and chena cultivation; third, the sickly nature of the population; and fourth and principally, the failure of the coffee crop.[94]

To make very clear his disagreement with Dickson, he asserted that it was the 'harrying of the people for their arrears of tax that was the cause of their ultimate widespread misery and distress.'[95] While officials could disagree over the reasons for destitution, what was not at issue was the fact that between 1880 and 1892, 9.4 per cent of all the paddy land on the island had been sold off by the government. In Badulla District the rate was 30.4 per cent.[96] What was not directly admitted was that a taxation policy had been created around a set of environmental expectations that were dashed by disease.

---

91 Wickremeratne, 'Grain consumption', 45.

92 Davis, *Late Victorian*, shows in detail how similar conjunctures of events produced widespread famines in Madras during these same years.

93 *SP*, 1889, 1–9.

94 Throughout the plantation areas, peasants complained of the effect of deforestation on the flow of water to their paddy fields on the valley bottoms. The AGA of Nuwara Eliya District in 1884 wrote in his Annual Report that 'There can be no doubt, I think, that the clearing of almost all the high slopes at the head of this valley as coffee estates, ... has seriously affected the water supply of the fields below and disregard or at least ignorance of the probable result was shown when so much forest was sold to be opened as estates.' *ARC*, 1884, 66a.

95 *SP*, 1889, 3, 19.

96 M. Roberts, 'Some Comments on Ameer Ali's Paper', *Ceylon Studies Seminar* 3b (1970/72), 20. There were no figures available for Nuwara Eliya District, one of the hardest hit because there was no accurate record of the amount of land under paddy cultivation.

In 1892 newly-arrived Governor Havelock abolished the Paddy Tax. Although this entailed an annual loss to the government of 500,000 rupees, the sum was recouped by the imposition of new taxes such as stamp and salt taxes and import duties, which predictably fell disproportionately on the peasantry. Roberts argues that whilst the Paddy Tax was seen as abhorrent *in principle* by all, for the British had come to see food taxes in this way, that it was supported *in practice* by some because the 'East' was seen as 'the great exception', the special case, where such taxes served to stir peasants from their Oriental lethargy. Others supported the tax on the grounds of expediency; the means justifying the ends of meeting the financial needs of the colony.[97] In both cases, these food taxes were seen as 'a necessary evil.' Suffice it to say that the notion of 'necessary evil' is morally repellent, as it can serve to justify nearly anything.[98]

Conclusion

In this chapter I have traced the trajectory of a disease of the coffee plant that within a little over a decade destroyed an industry that had been the life-blood of Ceylon for 40 years. The government marshalled as much knowledge and technical expertise as possible to cure the disease to no avail, and then switched its support to other plantation crops. The government was less successful at reducing the impact of the failure of coffee on the non-European population. Attempts at establishing new regimes of governmentality in regard to the care and control of estate labour were as usual fended off by planters in the name of liberalism and the interests of the colony's economic foundation. Worse than nothing was done for the peasantry. Portions of the highlands became zones of social abandonment, with the hungry poor in some of the planting districts sacrificed by the government in order to make up the financial shortfall caused by the disease.

---

97 Roberts, 'Grain taxes', 142.

98 Legg, *Spaces of Colonialism*, Ch. 3, discusses this notion in regard to the suppression of crime in India.

Chapter 8

Conclusion

> [i]n thinking of the mechanisms of power, I am thinking ... of its capillary forms of existence, the point where power reaches into the very grain of individuals, touches their bodies and inserts itself into their actions and attitudes, their discourses, learning processes and everyday lives.
>
> Michel Foucault[1]

The above quotation describes in brief what Foucault terms the 'synaptic regimes of power,' which operate through the body as well as the mind.[2] This book has documented the profound challenges in establishing such a regime of power in mid-nineteenth century Ceylon. I have explored a range of regimes of practice – labour, health, policing – and in spite of repeated attempts by the British to introduce current European values of conduct through these practices into the very grain, the very bodies, of the estate labourers and Sinhalese peasantry, we have seen how strenuously this was resisted. The Tamil labourers evaded many of the disciplinary practices set up to maximize efficiency and profit on the estates. Labourers and peasants stole coffee and cattle were allowed to trespass on estates where they trampled crops. Labourers hid their diseases as best they could from the medical police and the planters. They died on estates and roads rather than go to hospital. The lack of British legitimacy was such that there appeared to be little or no stigma attached to the evasion of colonial regulation. Such evasion and refusal to accept British codes of conduct were undoubtedly the most effective weapons of the weak. The resistance was very quiet. Its power lay in the fact that it was leaderless and dispersed; it was virtually everywhere and nearly impossible to combat. It was the will to survive and not British power that entered into the very grain of individuals.

The responses of the British to such strong resistance – to what the British (mistakenly) saw as a virtual imperviousness to British cultural initiatives – was to target power at a more basic bodily level. Where the inculcation of values failed, the British administrators and planters resigned themselves to coercion. They aimed their programmes and policies at the peasants' and especially the migrant labourers' very bodily existence – their aim was to 'make survive' and to attempt to squeeze out every ounce of labour from their bodies. Governmentality became reluctantly authoritarian. The ambitions of modern governmentality (the conduct of conduct, as Foucault calls it) competed relatively unsuccessfully with the older mercantilist principles – the extraction of wealth as the highest priority. The failure of liberalism to produce docile bodies came at a cost. Massive surveillance, coercion and corporal

---

1   Foucault, 'Prison Talk', 39.
2   Ibid., 39.

punishment of the peasantry and labourers were increasingly resorted to. Even the British planters – who were more receptive to new ideas flowing from Britain – even they, at times, rejected strict self-discipline and control over their passions. As we have seen in Chapter 3, this partial rejection of British norms of middle class conduct on the part of planters came at a cost – the loss of their full racial identity. And for the 'people in between' – the Burghers and second-generation tropical-English, who were suspended between Europe and Asia – the long shadow of racism and environmental determinism hung darkly over them, no matter how they conducted themselves. Disciplinary power might penetrate their bodily practices, but the discourses of race and the tropics insured that bodily practices did not render them fully European either in their own estimation or in the eyes of others.

We have seen how planters were 'men on the spot' facing their own fears of the tropics, tropical disease and tropical peoples. They resisted ideas of humane governmentality emanating from Britain, and fought tooth and nail to oppose costly programmes aimed at managing the estate labourers' welfare. Operating under the banner of laissez-faire liberalism, they largely rejected governmentality in favour of authoritarian biopower which at best sought to merely 'make survive.' On failing estates in the tropical disease environment of the hills, the line between merely 'making survive' and the sovereign decision to 'allow to die' was very fine indeed. In the face of government opposition or sometimes with grudging government support, planters employed colonialist mythologies of the tropics and racial difference, backed (albeit uncertainly) by prevailing biological and social scientific theories, to justify their persistent undermining of government policy. Despite all their efforts and exploitative practices, the planters failed.

Although the effects of British colonialism are evident in Ceylon today and a plantation economy still exists there, the British colonial dreams in Ceylon caused more misery and ecological havoc than could have been predicted by nineteenth century science. A conventional view might see the coffee period as constituting a particular social and economic order that broke down under the impact of the coffee leaf disease and was followed by a period of relative disorder as the projects of the European planters, labourers and peasants failed over a relatively short period of time. But what I have tried to demonstrate is that the ordering that did exist throughout the coffee period was tenuous at best. It constituted, to use Law's terms, only 'small pools of ordering'[3] and these were in constant danger of evaporating under the tropical sun. Whether one looks at labour, or property, or medicine, order was far more theoretical than it was ever achieved in practice. What ordering did exist had to be continually renegotiated and compromised. It was partial, contingent, temporary, one could even say evanescent. It was, to use Foucault's phrase, 'the glitter above the abyss.'[4]

Modernity was never a purely European project, but necessarily locally differentiated, fractured and negotiated–in fact, it makes little sense to speak of modernity as a unified project or even a discourse or regime of practices. As Sherry Ortner has stated, 'pieces of reality however much borrowed from or imposed on others, are woven together through the logic of a group's own locally and historically

---

3   J. Law, *Organizing Modernity* (Oxford: Blackwell, 1994), 5.
4   M. Foucault, *The Order of Things* (New York: Random House, 1970), 251

evolved *bricolage*.[5] Colonial modernity was not the tropicalization of its western form; it was not the violation of metropolitan norms. Modernity and its various manifestations such as governmentality (in all its various manifestations) was always multiple and spatially differentiated. Knowledge practices and techniques flowed into the colonies and were transformed there not as a new imposed order but as newly evolving local and extra-local forms.

I have explored the way the coffee labourers, and plantation economies were affected by local conditions in Ceylon, by prosperity, famine and epidemics in southern India, by growing conditions in other coffee-producing regions like Brazil and Java, by depressions in Europe and shifting demands and shifting world prices for their products. Throughout all of this, some planters made money and others failed. The ordering, such as it was, was both incredibly complex and tenuous. As the leaf disease progressively erased coffee as a viable economic enterprise in the highlands of Ceylon, nearly everyone failed. But this new disease-led failure, was evanescent as well, because some planters left the island, or died, while others shifted production into other crops such as cinchona, cacao, or tea. Some labourers died, some went back to India and others stayed on to work the new tea estates. And new pools of diverse and interacting practical orderings were set in motion; there arose new complexes 'composed of men and things' and new chains of events linking the fate of tropical Ceylon to events around the globe.

One of the important contributions of Foucault's notions of governmentality and biopower is the unsettling of the nature/culture divide. He argues that it is neither territory nor population alone that became the target of government:

> [w]hat government has to do with is not territory but, rather, a sort of complex composed of men and things. The things, in this sense, with which government is to be concerned are in fact men, but men in their relations, their links, their imbrication with those things that are wealth, resources, means of subsistence, the territory with its specific qualities, climate, irrigation, fertility, and so on.[6]

---

5   S. Ortner, 'Resistance and the Problem of Ethnographic Refusal', *Comparative Studies in Society and History* 37 (1995), 176.

6   Foucault, 'Governmentality', 93. Geographers G. O'Tuathail, *Critical Geopolitics: The Politics of Writing Global Space* (Minneapolis: University of Minnesota Press, 1996); B. Braun, 'Producing Vertical Territory: Geology and Governmentality in Late Victorian Canada', *Ecumene* 7 (2000): 7–45; and Rose-Redwood, 'Governmentality', 469–486 have questioned Foucault's view that modern governmentality's principal focus was population rather than territory. This quotation suggests that he saw both as highly intertwined. Nevertheless, in the case of Ceylon although a Survey Department was set up and great importance was placed on the legibility of the territory, in practice more resources were applied to managing the population, especially the immigrant workers. Ian Barrow points out:

'The Ceylon Survey Department was established in 1800, and by the end of the century it was one of the largest in the British Crown Colonies. Yet, the department conducted no systematic triangulation of the island and until the 1870s no regular topographical survey. Only one significant map of Ceylon was published as a result of modern surveying before the end of the nineteenth century'. (I.J. Barrow, 'Surveying in Ceylon during the Nineteenth Century', *Imago Mundi* 55 (2003), 81)

Despite the desire for knowledge of the territory as evidenced by the establishment of the Survey Department, governmental resources were stretched and, as we have seen repeatedly,

The deforestation of the Kandyan highlands and the monocultivation of coffee as an export product upon which the colony had come to depend caused the populations and ecologies of the region to become deeply entangled with many other places in the world. Colonial planters and government administrators, doctors and hospital administrators, and scientists all experimented with various techniques of discipline and biopower. They attempted to visualize, predict and control these complex imbrications of human populations and nature and their extensive spatial linkages. They attempted to understand the difference that tropical nature, climate and race made to the way these could and should be governed. However, these complexities escaped them both theoretically and practically. Coffee eventually collapsed and the primary agent was an untreatable fungus, *Hemileia vastatrix*. Arriving unpredictably, in a manner still not known with any certainty, in a region approaching a condition of near monocultivation, the fungus spread rapidly and uncontrollably.

The global networks that initiated the plantation agricultural practices – and then the disease which threatened the industry – were in turn drawn upon in the search for a cure. The government of Ceylon marshalled its bio-political power to organize a response to this threat to the economic viability of the colony. Networks of planters and AGAs were employed to survey the disease from year to year around the island. The Botanic Garden at Peradeniya conducted research, recruiting scientific knowledge from Kew and importing other strains of coffee from botanic gardens around the empire. And yet, the disease resisted all of these techniques of governmentality. The sudden rise of *Hemileia vastatrix* in the highlands of Ceylon in 1869 and its wildfire spread across the island was unquestionably an artefact of nature/culture. We saw in Chapter 7 how the starvation of Sinhalese peasants, the grinding poverty of Tamil estate labourers and despair of bankrupt planters was caused by the radically transformed ecology of the highlands and a fungus of the coffee plant which spread uncontrollably due to decisions made both locally and in far away places; by earlier shifts in local tax policy that operated on unsubstantiated environmental expectations regarding peasant and colonialist coffee production; by the unforeseen effects of treating plantations as abstract spaces as if they were isolated from the surrounding ecologies, the inadvertent introduction of a new pest while seeking a cure for the fungus; by shifting financial conditions in Europe; and by conditions of production in other nodes of the network such as Brazil. There can be no easy distinction between nature and culture or the local and the global in these long networks of causation. Rather the question is one of exploring how nature/culture interpenetrate and how various actors and localities become vulnerable and also complicit in a network in ways that are so complex that they escaped governmental or scientific management at the time and are only slightly less difficult to visualize in retrospect.

---

the focus of the state was guided in large part by the demands of the planting community. The result was that knowledge and understanding of the land and people of Ceylon was highly uneven. As Barrow states in answer to the question of why so little topographical and trigonometrical surveying was carried out: 'The Survey Department was poorly organized and the surveyors were preoccupied with road construction and the mapping of Crown land for sale as coffee estates.' Barrow, 'Surveying in Ceylon', 81.

Bibliography

Adas, M. 'From Foot Dragging to Flight: The Evasive History of Peasant Avoidance Protest in South and Southeast Asia', *Journal of Peasant Studies* 13 (1986): 64–86.

Adas, M. *Machines as the Measure of Men: Science, Technology, and Ideologies of Western Dominance* (Ithaca: Cornell University Press, 1989).

Addresses Delivered in the Legislative Council Vol. 2. (Colombo: Government Printer, 1900).

Administration Reports, Ceylon. (*ARC*) (Colombo: Government Printer).

Agamben, G. *Remnants of Auschwitz: The Witness and the Archive* (New York: Zone Books, 1999).

Agnew, J. and Coleman, M. 'The Problem With Empire', in *Space, Knowledge and Power: Foucault and Geography*, edited by J. Crampton and S. Elden (Aldershot: Ashgate, 2007), 17–39.

Alloula, M. *The Colonial Harem* (Minneapolis: University of Minnesota Press. 1986).

Ameer Ali, A.C.M. 'Peasant Coffee in Ceylon During the Nineteenth Century', *Ceylon Journal of History and Social Science* 2 (1972): 50–59.

Ameer Ali, A.C.M. 'Rice and Irrigation in the Twentieth Century', *The Ceylon Historical Journal* 25 (1978): 26–41.

Amin, S. *Event, Metaphor, Memory: Chauri Chaura 1922–1992* (Delhi: Oxford University Press, 1995).

Anderson, B. *Imagined Communities* (London: Verso, 1983).

Annesley, J. *Researches into the Causes, Nature, and Treatment of the More Prevalent Diseases of India and the Warm Climates Generally* (London: Longman, Brown, Green and Longman, 1841).

Anonymous *Days of Old: Or the Commencement of the Coffee Enterprise in Ceylon by Two of the Pioneers* (Colombo: A.M. and J. Ferguson, 1878).

Anonymous, *Fickle Fortune in Ceylon* (Madras: Addison and Company, 1887).

A Planter (Richard Wade Jenkins) *Ceylon in the Fifties and the Eighties, a Retrospect and Contrast of the Vicissitudes of the Planting Enterprise During a Period of Thirty Years and of Life and Work in Ceylon* (Colombo: A.M. and J. Ferguson, 1886).

Arasaratnam, C.T. 'A Brief History of the Development of Labour Relations in Ceylon', *Ceylon Labour Gazette* (April, 1970).

Arasaratnam, S. *Ceylon* (Englewood Cliffs, New Jersey: Prentice Hall, 1964).

Armstrong, D. 'Public Health Spaces and the Fabrication of Identity', *Sociology* 27 (1993): 393–410.

Arnold, D. *Colonizing the Body: State Medicine and Epidemic Disease in Nineteenth Century India* (Berkeley: University of California Press, 1993).

Arnold, D. *The Problem of Nature: Environment, Culture and European Expansion* (Oxford: Blackwell, 1996).

Asad, T. 'Conscripts of Western Civilization', in *Dialectical Anthropology: Essays in Honor of Stanley Diamond, Vol. 1 Civilization in Crisis*, edited by C. Gailey. (Gainsville: University Press of Florida, 1992).

Baber, Z. *The Science of Empire: Scientific Knowledge, Civilization and Colonial Rule in India* (Albany: SUNY Press, 1996).

Baker, S.W. *Eight Years in Ceylon* (London: Longmans, Green and Co, 1855).

Ballhatchet, K. *Race, Sex and Class Under the Raj* (London: Weidenfeld and Nicolson, 1980).

Bandarage, A. *Colonialism in Sri Lanka: The Political Economy of the Kandyan Highlands 1833–1886* (Berlin: Mouton, 1983).

Barnett, C. 'Sing Along with the Common People: Politics, Postcolonialism and Other Figures', *Environment and Planning D. Society and Space* 15 (1997): 137–54.

Barron, T.J. 'Science and the Nineteenth Century Ceylon Coffee Planters', *The Journal of Imperial and Commonwealth History* 16 (1987): 5–21.

Barrow, I.J. 'Surveying in Ceylon During the Nineteenth Century', *Imago Mundi* 55 (2003): 81–96.

Barthes, R. *Mythologies* (London: Jonathan Cape, 1972).

Bashford, A. *Purity and Pollution: Gender, Embodiment and Victorian Medicine* (London: Macmillan, 1998).

Bastiampillai, B. 'The South Indian Immigrants' Trek to Ceylon in the Mid-Nineteenth Century', *Sri Lankan Journal of Social Science* 7 (1984): 41–66.

Bastiampillai, B. 'The Administration of Sir William Gregory. Governor of Ceylon 1872–1877', *The Ceylon Historical Journal* 12 (Dehiwala: Tisara Prakasakayo, 1968).

Behal, R.P. 'Forms of Labour Protest in the Assam Tea Plantations, 1900–1930', *Economic and Political Weekly* 20 (1985): 19–26.

Behal, R.P. and P. Mohapatra, 'Tea and Money Versus Human Life: the Rise and Fall of the Indenture System in the Assam Tea Plantations, 1840–1908', *The Journal of Peasant Studies* 19 (1992): 142–72.

Bennett J.W. *Ceylon and its Capabilities* (London: W.H. Allen, 1843).

Bewell, A. *Romanticism and Colonial Disease* (Baltimore: Johns Hopkins University Press, 1999).

Bhatia, B.M. *Famines in India: 1860–1965* (London: Asia House Publishing, 1963).

Biehl, J. 'Vita: Life in a Zone of Social Abandonment', *Social Text* 68 (2001): 131–49.

Blume, M. *Côte d'Azur: Inventing the French Riviera* (London: Thames and Hudson, 1992).

Blunt, A. *Domicile and Diaspora: Anglo-Indian Women and the Spatial Politics of Home* (Oxford: Blackwell, 2005).

Bowd, G. and D. Clayton, 'Fieldwork and Tropicality in French Indochina: Reflections on Pierre Gourou's "Les paysans du Delta Tonkinois, 1937"', *Singapore Journal of Tropical Geography* 41 (2003): 147–68.

Boyd, W. 'Autobiography of a Periya Durai', *Ceylon Literary Register* 3 (1888).

Boyd, W. 'Ceylon and its Pioneers', *Ceylon Literary Register* 2 (1888).

Braun, B. 'Producing Vertical Territory: Geology and Governmentality in late Victorian Canada', *Ecumene* 7 (2000): 7–45.

Briggs, C.L. and C. Mantini-Briggs, *Stories in the Time of Cholera: Racial Profiling During a Medical Nightmare* (Berkeley: University of California Press, 2003).

British Parliamentary Papers (BPP).

Brown, A. *The Coffee Planter's Manual* (Colombo: Ceylon Observer Press, 1880).

Brown, S. *Life in the Jungle, or Letters from a Planter to his Cousin in London* (Colombo: Herald Press, 1845).

Burton, R. *Goa and the Blue Mountains: Or, Six Months of Sick Leave* (London: R. Bentley, 1851).

Butler, J. 'Performativity's Social Magic', in *The Social and Political Body*, edited by T.R. Schatzki and W. Natter (New York: Guilford, 1996), 5–38.

Campbell, C. *Philloxera* (London: Harper Collins, 2004).

Capper, J. *Old Ceylon:Sketches of Ceylon in Olden Times* (London: W.B. Whittington, 1878).

Carpen, *The Diary of a Kangany* (Colombo: Privately Published, No Date).

Carpenter, E. *From Adam's Peak to Elephanta* (London: Kessinger, 1910).

Carter, M. 'Strategies of labour mobilisation in colonial India: the recruitment of indentured workers for Mauritius', in *Plantations, Proletarians and Peasants in Colonial Asia*, edited by E.V. Daniel, H. Bernstein, and T. Brass (London: Routledge, 1992), 229–45.

Carter, M. *Servants, Sirdars and Settlers: Indians in Mauritius 1843–1874* (Delhi: Oxford University Press, 1995).

Cell, J. 'The Imperial Conscience', in *The Conscience of the Victorian State*, edited by P. Marsh (Syracuse: Syracuse University Press, 1979), 173–213.

The Ceylon Civil Service Manual (Colombo: F. Fonseka, Printer, 1872).

Ceylon Examiner.

Ceylon Hansard, Debates of the Ceylon Legislative Council. Colombo: Government Printer.

Ceylon Observer.

Ceylon Plantation Gazetteer (Colombo).

Chadwick, E. *Report on the Sanitary Condition of the Labouring Population of Gt Britain* (Edinburgh: Edinburgh University Press, 1965 [1842]).

Chattergee, P. 'Two Poets and Death: On Civil and Political Society in the Non-Christian World', in *Questions of Modernity*, edited by T. Mitchell (Minneapolis: University of Minnesota Press, 2000), 35–48.

Chattopadhyaya, H. *Indians in Sri Lanka* (Calcutta: OPS Publishers, 1979).

Clarence-Smith, W. G. 'Planters and Small-Holders in Portuguese Timor in the Nineteenth and Twentieth Centuries', *Indonesia Circle* 57 (1992): 15–30.

Clarence-Smith, W.G. 'The spread of coffee cultivation in Asia from the seventeenth to the early nineteenth century', in *Le Commerce du Café Avant L'etre des Plantations Coloniales*, edited by M. Tuchscherer (Cairo: Institut Français d'Archéologie Orientale, 2001), 371–84.

Clarence-Smith, W.G. 'The Coffee Crisis in Asia, Africa, and the Pacific, 1870–1914', in *The Global Coffee Economy in Africa, Asia and Latin America, 1500–1980*, edited by W.G. Clarence-Smith and S. Topik (Cambridge: Cambridge University Press, 2003), 100–19.

Clayton, D. *Islands of Truth: The Imperial Fashioning of Vancouver Island* (Vancouver: University of British Columbia Press, 2000).

Clayton, D. 'Imperial Geographies', in *A Companion to Cultural Geography*, edited by J.S. Duncan, N.C. Johnson and R.H. Schein (Oxford: Blackwell, 2004), 449–68.

Cleary, M. 'Land Codes and the State in French Cochinchina, c. 1900–1940', *Journal of Historical Geography* 29 (2003): 356–75.

Coderington, H.W. *Ancient Land Tenure and Revenue in Ceylon* (Colombo: Ceylon Government Press, 1938).

Cohn, B. *Colonialism and its Forms of Knowledge: The British in India* (Princeton: Princeton University Press, 1996).

Colonial Office (CO).

Collingham, E.M. *Imperial Bodies* (Cambridge: Polity, 2001).

Comaroff, J.L. and Comaroff, J, *Ethnography and the Historical Imagination* (Boulder: Westview, 1992).

Connerton, P. *How Societies Remember* (Cambridge: Cambridge University Press, 1989).

Courtenay, P.P. *Plantation Agriculture* London: (Bell and Hyman, 1980).

Crampton, J. and S. Elden (eds), *Space, Knowledge, and Power: Foucault and Geography* (Aldershot: Ashgate, 2007).

Cullimore, D.H. *The Book of Climates: Acclimatisation; Climatic Disease; Health Resorts and Mineral Springs; Sea Sickness; and Sea Bathing* (London: Bentley, 1890).

Curtin, P.D. *Death By Migration: Europe's Encounter with the Tropical World in the Nineteenth Century* (Cambridge: Cambridge University Press, 1989).

Daniell, L.W. *Wentworth-Reeve Papers* (Centre for South Asian Studies, Cambridge University, 1866).

Das, V. 'Subaltern as perspective', in *Subaltern Studies VI: Writings on South Asian History and Society*, edited by R. Guja (New Delhi: Oxford University Press), 310–24.

Davidson, A. *Hygiene and Diseases of Warm Climates* (Edinburgh: Pentland, 1893).

Davis, M. *Late Victorian Holocausts* (London: Verso, 2001).

Dean, M. *Governmentality: Power and Rule in Modern Society* (London: Sage, 1999).

De Butts, L. *Rambles in Ceylon* (London: W.H. Allen, 1841).

De Certeau, M. *The Practice of Everyday Life.* (Berkeley: University of California Press, 1984).

Dep, A.C. *A History of the Ceylon Police 1866–1913* (Colombo: Police Amenities Fund, 1969).

De Silva, C.R. *Ceylon Under the British Occupation 1795–1833 Volume 2* (Colombo: The Colombo Apothecaries' Company, 1962).

De Silva, K.M. *Social Policy and Missionary Organisations in Ceylon 1840 to 1855* (London: Longmans, 1965).

De Silva, K.M. 'Indian Immigration to Ceylon—the First Phase, c. 1840–1855', *Ceylon Journal of Historical and Social Studies* 4 (1961): 106–37.

De Silva, K.M. (ed.), *Letters on Ceylon 1846–50. The Administration of Viscount Torrington and the Rebellion of 1848* (Kandy: K.G.V. De Silva and Sons, 1965).

De Silva, K.M. 'Resistance Movements in Nineteenth Century Sri Lanka', in *Collective Identities, Nationalisms and Protest in Modern Sri Lanka*, edited by M. Roberts (Colombo:Marga, 1979), 129–53.

De Silva, S.B.D. *The Political Economy of Underdevelopment* (London: Routledge and Kegan Paul, 1982).

Digby, W. *Life of Sir Richard Morgan* Vol. 2. (Madras: Addison and Company, 1879)

Dirks, N.B. *Castes of Mind: Colonialism and the Making of Modern India* (Princeton: Princeton University Press, 2001).

Driver, F. 'Imagining the Tropics: Views and Visions of the Tropical World', *Singapore Journal of Tropical Geography* 25 (2004): 1–17.

Driver, F. and Martins, L. 'Uses and Visions of the Tropical World', in *Tropical Visions in an Age of Empire*, edited by F. Driver and L. Martins (Chicago: University of Chicago Press, 2005), 3–20.

Driver, F. and B. Yeoh, 'Constructing the Tropics: Introduction.' *Singapore Journal of Tropical Geography* 21 (2000): 1–5.

Duncan, J.S. *The City as Text: The Politics of Landscape Interpretation in the Kandyan Kingdom* (Cambridge: Cambridge University Press, 2005).

Duncan, J.S. 'Home alone? Masculinity, Discipline and Erasure in Mid-Nineteenth Century Ceylon', in *Gendered Landscapes*, edited by L. Dowler et al. (New York: Routledge, 2005), 19–33.

Duncan J.S. 'Sombres Pensées Dans la Maison Coloniale: Masculinité, Contrôle et Refoulement Domestiques à Ceylan au Milieu du XIXème Siècle', in *Espaces Domestiques. Construire, Aménager, Représenter*, edited by B. Collignon and J.-F. Staszak, 341–53. (Paris: Bréal, 2004).

Duncan, J.S. 'Complicity and Resistance in the Colonial Archive: Some Issues of Method and Theory in Historical Geography', *Historical Geography* 27 (1999): 119–128.

Duncan, J.S. 'The Struggle to be Temperate: Climate and "Moral Masculinity" in Mid-Nineteenth Century Ceylon', *Singapore Journal of Tropical Geography* 21 (2000): 34–47.

Duncan, J. 'Embodying Colonialism?: Domination and Resistance in 19th Century Ceylonese Coffee Plantations', *Journal of Historical Geography* 28 (2002): 317–38.

Edmond, R. 'Returning Fears: Tropical Disease and the Metropolis', in *Tropical Visions in an Age of Empire*, edited by F. Driver and L. Martins (Chicago: University of Chicago Press, 2005), 175–94.

Edney, M. *Mapping an Empire: The Geographical Construction of British India, 1765–1843* (Chicago: University of Chicago Press, 1997).

Engels, F. *The Condition of the Working Class in England* (Oxford: Blackwell, 1958).

Evans, R.J. *Death in Hamburg: Society and Politics in the Cholera Years, 1830–1910* (Oxford: Clarendon, 1987).

Evans, R.J. 'Epidemics and Revolutions: Cholera in Nineteenth Century Europe', in *Epidemics and Ideas: Essays on the Historical Perception of Pestilence*, edited by T. Ranger and P. Slack (Cambridge: Cambridge University Press, 1992), 149–74.

Eyler, J.M. 'The Sick Poor and the State: Arthur Newsholme on Poverty, Disease, and Responsibility', in *Framing Disease*, edited by C.E. Rosenberg and J. Golden (New Brunswick: Rutgers University Press, 1992), 275–96.

Ferguson, A.M. and J. Ferguson, *The Ceylon Directory for 1874* (Colombo: A.M. and J. Ferguson, 1874).

Ferguson, A.M. *Summary of Information Regarding Ceylon* (Colombo: A.M. and J. Ferguson, no date).

Ferguson, A.M. *Ceylon. Summary of Useful Information* (Colombo: A.M. and J. Ferguson, 1859).

Ferguson's Ceylon Directory (Colombo: A.M. and J. Ferguson, 1865–66).

Ferguson, J. *Ceylon in the Jubilee Year* (London: J. Haddon and Company, 1887).

Ferguson, J. *Ceylon in 1893* (London: Sampson, Low, Marston, Searle and Rivington, 1893).

Ferguson, J. 'The Prospects of England's Chief Colony. An Interview with a Ceylon Journalist (Mr. J. Ferguson). *The Pall Mall Gazette* August 29, 1884', in *Ceylon and Her Planting Enterprise: in Tea, Cacao, Cardomom, Chinchona, Coconut and Areca Palms. A Field for the Investment of British Capital and Energy*, edited by A.M Ferguson and J. Ferguson. (Colombo: A.M. and J. Ferguson, 1885).

Ferguson, J. *Pioneers of the Planting Enterprise in Ceylon. Vol. 1.* (Colombo: A.M. and J. Ferguson. 1894).

Fernando, M.R. and W. O'Malley, 'Peasants and Coffee Cultivation in Cirebon Residency', in *Indonesian Economic History in the Dutch Colonial Era*, edited by A. Booth et al. (New Haven: Yale University Press, 1990), 171–86.

Forbes, M. *Eleven Years in Ceylon, Comprising Sketches of the Field Sports and the Natural History of that Colony, and an Account of its History and Antiquities*. 2 Vols. (London: Richard Bentley, 1840).

Forrest, D.M. *A Hundred Years of Ceylon Tea, 1867–1967*. (London: Chatto and Windus, 1967).

Forrest, D. 'Hundred Years of Achievement. *The Times of Ceylon Tea Centenary Supplement* (31 July 1969).

Foucault, M. *The Order of Things* (New York: Random House, 1970).

Foucault, M 'Docile Bodies', in *Discipline and Punish: The Birth of the Prison* (New York: Vintage, 1979), 135–69.

Foucault, M. *Discipline and Punish: The Birth of the Prison* (New York: Vintage, 1979).

Foucault, M. 'The Eye of Power', in *Power/Knowledge: Selected Interviews and Other Writings, 1972–1977*, edited by C. Gordon (New York: Pantheon, 1989), 146–65.

Foucault, M. 'Prison Talk', in *Power/Knowledge: Selected Interviews and Other Writings, 1972–1977*, edited by C. Gordon (New York: Pantheon, 1980), 37–54.

Foucault, M. *The History of Sexuality: An Introduction. Volume 1.* (New York: Vintage, 1990).

Foucault, M. 'Governmentality', in *The Foucault Effect: Studies in Governmentality*, edited by G. Burchell, C. Gordon and P. Miller (Chicago: University of Chicago Press, 1991), 87–104.

Foucault, M. *The Essential Works of Foucault 1954–1984; Power* (Penguin: London, 2001[1978]).

Foucault, M. *The Essential Foucault*, edited by N. Rose (New York: New Press, 2003).

Gaskell, P. *The Manufacturing Population of England, Its Moral, Social and Physical Conditions, and the Changes Which Have Arisen from the Use of Steam Machinery* (London: Baldwin and Cradock, 1833).

Gay, P. *Pleasure Wars: The Bourgeois Experience. Victoria to Freud. Volume 5* (London: Fontana, 1998).

Genovese, M. *Roll Jordan Roll: The World the Slaves Made* (New York: Random House, 1972).

Gikandi, S. *Maps of Englishness: Writing Identity in the Culture of Colonialism* (New York: Columbia University Press, 1996).

Gilbert, P.K. *Mapping the Victorian Social Body* (Albany: State University of New York Press, 2004).

Glacken, C. *Traces on the Rhodian Shore* (Berkeley: University of California, 1967).

Gough, K. *Rural Society in Southeast India* (Cambridge: Cambridge University Press, 1981).

Gregor, H. 'The Changing Plantation', *Annals of the Association of American Geographers*, 55 (1965): 221–238.

Guja, R. *Elementary Aspects of Peasant Insurgency in Colonial India* (Delhi: Oxford University Press, 1992).

Hacking, I. *The Taming of Chance* (Cambridge: Cambridge University Press, 1990).

Hall, C. *White, Male and Middle Class: Explorations in Feminism and History* (Cambridge: Polity Press, 1992).

Hall, C. *Civilising Subjects: Metropole and Colony in the English Imagination, 1830–1867* (Cambridge: Polity, 2002).

Hall, D. *Free Jamaica, 1838–1865: An Economic History* (New Haven: Yale University Press, 1959).

Haller, J.S. *Outcasts From Evolution: Scientific Attitudes of Racial Inferiority 1859–1900* (Carbondale: Southern Illinois University Press, 1995).

Hamilton, V.M. and S.M. Fasson, *Scenes in Ceylon* (H.W. and A.W. Cave, 1881).

Hannah, M. *Governmentality and the Mastery of Territory in Nineteenth-Century America* (Cambridge: Cambridge University Press, 2000).

Harrison, M. *Public Health in British India: Anglo-Indian Preventive Medicine, 1859–1914* (Cambridge, Cambridge University Press, 1994).

Harrison, M. 'A Question of Locality: The Identity of Cholera in British India, 1860–1890', in *Warm Climates and Western Medicine: The Emergence of Tropical Medicine, 1500–1900*, edited by D. Arnold (Amsterdam: Rodopi, 1996), 133–59.

Harrison, M. *Climates and Constitutions: Health, Race, Environment and British Imperialism in India, 1600–1850* (New Delhi: Oxford University Press, 1999).

Harrison, M. *Disease and the Modern World* (Cambridge: Polity, 2004).

Heidemann, F. *Kanganies in Sri Lanka and Malaysia: Tamil Recruiter-cum-Foreman as a Sociological Category* (Munich: Anacon, 1992).

Hepworth, M. 'Privacy, Security and Respectability: The Ideal Victorian Home', in *Ideal Homes?* edited by T. Chapman and J. Hockey (London: Routledge, 1999), 17–29.

Himmelfarb, G. *The Idea of Poverty: England in the Early Industrial Age* (New York: Knopf, 1984).

Houlgate, S. 'Vision, Reflection, and Openness', in *Modernity and the Hegemony of Vision*, edited by D. Levin (Berkeley: University of California Press, 1993), 87–123.

Howell, P. 'Foucault, Sexuality, Geography', in *Space, Knowledge and Power: Foucault and Geography* edited by J. Crampton and S. Elden (Aldershot: Hampshire, 2007), 291–316.

Hunt, J. 'On Ethno-Climatology; or the Acclimatization of Man', *Transactions of the Ethnological Society of London* n.s.2 (1863).

Hyam, R. *Empire and Sexuality: The British Experience* (Manchester: Manchester University, 1990).

Illustrated London News.

Iyer, R. *Utilitarianism and All That* (London: Concord Grove Press, 1983).

Jayawardene, K.V. *The Rise of the Labour Movement in Ceylon* (Durham: Duke University Press, 1972).

Jayawardena, K. *Nobodies to Somebodies: The Rise of the Colonial Bourgeoisie in Sri Lanka* (Colombo: The Social Scientists' Association, 2000).

Jayawardene, L. R. *The Supply of Sinhalese Labour to Ceylon Plantations (1830–1930): A Study of Imperial Policy in a Peasant Society* (Ph.D. Thesis, University of Cambridge, 1963).

Jeffreys, J. *The British Army in India: Its Preservation by an Appropriate Clothing, Housing, Location, Recreative Employment, and Hopeful Encouragement of the Troops* (London: Adamant, 2005 [1858]).

Jepson, W. 'Of Soil, Situation and Salubrity: Medical Topography and Medical Officers in Nineteenth Century British India', *Historical Geography*, 32 (2004): 137–55.

Johnson, J. *The Influence of Tropical Climates, More Especially of the Climate of India, On European Constitutions; and the Principal Effects and Diseases Thereby Induced, their Prevention and Removal, and the Means of Preserving Health in Hot Climates Rendered Obvious to Europeans of Every Capacity.* 2nd Edn (London: B. & T. Kite, 1815).

Johnson, J. and Martin, J. *The Influence of Tropical Climates on European Constitutions* (London: S. Highley, 1846).

Kennedy, D. 'The Perils of the Mid Day Sun: Climatic Anxieties in the Colonial Tropics', in *Imperialism and the Natural World.* edited by J.M. MacKenzie (Manchester: Manchester University, 1990), 118–40.

Kennedy, D. *The Magic Mountains: Hill Stations and the British Raj* (Berkeley: University of California Press, 1996).

Kenny, J. T. 'Climate, Race, and Imperial Authority: the Symbolic Landscape of the British Hill Station in India', *Annals*, Association of American Geographers, 85 (1995): 694–714.

Kenny J. T. 'Claiming the High Ground: Theories of Imperial Authority and the British Hill Stations in India', *Political Geography*, 16 (1997): 117–39.

Kern, K. 'Gray Matters: Brains, Identities and Natural Rights', in *The Social and Political Body*, edited by T. Schatzki and W. Natter (London: Guilford, 1996), 103–21.

King, A.D. *Urban Colonial Development* (London: Routledge, 1976).

King, A.D. *The Bungalow: The Production of a Global Culture* (Oxford: Oxford University Press, 1995).

Kondapi, C. *Indians Overseas 1833–1949* (New Delhi: Oxford University Press, 1951).

Kumar, D. *Land and Caste in South India: Agricultural Labour in Madras Presidency during the Nineteenth Century.*(Cambridge: Cambridge University Press, 1965).

Kumar, M.S. 'The Evolution of the Spatial Ordering of Colonial Madras', in *Post-Colonial Geographies*, edited by A. Blunt and C. McEwan (New York: Continuum, 2002), 85–98.

Kurian, R. *State, Capital and Labour in the Plantation Industry in Sri Lanka 1834–1984* (Amsterdam: University of Amsterdam, 1989).

Kurian, R. 'Labour, Race and Gender on Coffee Plantations in Ceylon (Sri Lanka), 1934–1880', in *The Global Coffee Economy in Africa, Asia and Latin America, 1500–1980*, edited W.G. Clarence-Smith and S. Topik (Cambridge: Cambridge University Press, 2003), 173–90.

Kynsey, D. *Minutes of Evidence, Medical Aid Commission. Sessional Papers. Papers Laid Before the Legislative Council of Ceylon* (Colombo: Government Printer, 1869).

Lambert, D. *White Creole Culture, Politics and Identity During the Age of Abolition* (Cambridge: Cambridge University Press, 2005).

Lambert, D. and A. Lester. 'Geographies of Colonial Philanthropy', *Progress in Human Geography* 28 (2004): 320–41.

Landzelius, M. *personal communication.*

Law, J. *Organizing Modernity* (Oxford: Blackwell, 1994).

Lefebvre, H. *The Production of Space* (Oxford: Blackwell, 1991).

Legg, S. 'Foucault's Population Geography: Classifications, Biopolitical and Governmental Spaces', *Population, Space and Place* 2 (2005): 137–56.

Legg, S, 'Governmentality, Congestion, and Calculation in Colonial Delhi', *Social and Cultural Geography* 7 (2006): 709–29.

Legg, S. *Spaces of Colonialism: Discipline and Governmentality in Delhi, India's New Capital* (Oxford: Blackwell, 2007).

Lester, A. 'Constituting Colonial Discourse.' In *Post-Colonial Geographies,* edited by A. Blunt and C. McEwan (New York: Continuum, 2002), 29–45.

Lester, A. 'Obtaining the "Due Observance of Justice": The Geographies of Colonial Humanitarianism', *Environment and Planning: D* 20 (2002): 277–93.

Levin, D. (ed.) *Modernity and the Hegemony of Vision* (Berkeley: University of California Press, 1993).

Levine, P. *Prostitution, Race and Politics: Policing Venerial Disease in the British Empire* (New York: Routledge, 2003).

Lewis, F. *Sixty Four Years in Ceylon* (Colombo: Colombo Apothecaries, 1926).

Lewis, R.E. *Coffee Planting in Ceylon, Past and Present* (Colombo: Examiners Office, 1855).

Livingstone, D.N. 'The Moral Discourse of Climate: Historical Considerations on Race, Place and Virtue', *Journal of Historical Geography* 17 (1991): 413–34.

Livingstone, D.N. *The Geographical Tradition* (Oxford: Blackwell, 1992).

Livingstone, D.N. 'Tropical Climate and Moral Hygiene: the Anatomy of a Victorian Debate', *British Journal of the History of Science* 32 (1999): 93–110.

Livingstone, D.N. 'Race, Space and Moral Climatology: Notes Towards a Genealogy', *Journal of Historical Geography* 28 (2002): 159–80.

McCook, S. 'The Global Rust Belt: *Hemileia Vastatrix* and the Ecological Integration of World Coffee Production Since 1850', *Journal of Global History* 1 (2006): 177–95.

McEwan, C. 'Cutting Power Lines Within the Palace: Countering Paternity and Eurocentrism in the Geographical Tradition', *Transactions, Institute of British Geographers* 23 (1998): 371–84.

Mair, R.S. *A Medical Guide for Anglo-Indians. The European in India, or Anglo-Indian's Vade-Mecum. A Handbook of Useful and Practical Information for those Proceeding to or Residing in the East Indies* (London: King, 1871).

Mangan, J. R. *The Games Ethic and Imperialism* (Middlesex: Frank Cass, 1985).

Marby, H. *Tea in Ceylon* (Wiesbaden: F. Steiner Verlag, 1972).

Marshall, H. *Notes on the Medical Topography of the Interior of Ceylon* (London: Burgess and Hill, 1821).

Marshall, H. *Ceylon: A General Description of the Island and its Inhabitants* (London: H. Allen, 1846).

Martin, J.R. *The Influence of Tropical Climates on European Constitutions, Including Practical Observations on the Nature and Treatment of the Diseases of Europeans on Their Return from Tropical Climates* (London: Churchill, 1856).

Massey, D. *For Space* (London: Sage, 2005).

Mayhew, H. *London Labour and the London Poor. Volume 1.* (New York: A.M. Kelley, 1967 [1851]).

Mayhew, H. and J. Binny *The Criminal Prisons of London and Scenes of Prison Life* (London: Griffin, Bohn and Company, 1862).

Meegama, S.A. 'Cholera Epidemics and their Control in Ceylon', *Population Studies* 33 (1979): 143–56.

Mendis, G.C. ed. *The Colebrooke-Cameron Papers: Documents on British Colonial Policy in Ceylon 1796–1833* (Oxford: Oxford University Press, 1956).

Mendis, G.C., ed. 'Concessions for the Cultivation of Certain Agricultural Products, Regulation Number 4 of 21 September 1829', in *The Colebrooke Cameron Papers. Documents on British Colonial Policy in Ceylon 1796–1833. Volume 2*, edited by G.C. Mendis (Oxford: Oxford University Press, 1956), 279.

Metcalf, T.R. *Ideologies of the Raj* (Cambridge: Cambridge University Press, 1995).

Meyer, E. 'Between Village and Plantation: Sinhalese Estate Labour in British Ceylon', in *Asie du Sud: Traditions et Changements* (Paris: Colloques Internationaux du CNRS, 1979).

Meyer, E. 'Forests, Chena Cultivation, Plantations and the Colonial State in Ceylon 1840–1940', in *Nature and the Orient: The Environmental History of South and Southeast Asia*, edited by R.H. Grove, et. al., 793–827. (Delhi: Oxford University Press, 1998).

Meyer, E. 'Enclave Plantations, Hemmed-in Villages and Dualistic Representations in Colonial Ceylon', *Journal of Peasant Studies* 19 (1992): 199–228.

Meyer, E. 'The Plantation System and Village Structure in Rural South Asia', in *Rural South Asia: Linkages, Change and Development*, edited by P. Robb (London: Curzon, 1983).

Millie, P.D. *Thirty Years Ago: or Reminiscences of the Early Days of Coffee Planting in Ceylon*, Reprinted from the *Ceylon Observer* (Colombo: A.M. and J. Ferguson, 1878), 23–56.

Mills, L.A. *Ceylon under British Rule, 1795–1932* (London: Oxford University Press, 1933).

Mitchell, D. *The Lie of the Land* (Minneapolis: University of Minnesota Press, 1996).

Mitchell, T. 'The Stage of Modernity', in *Questions of Modernity*, edited by T. Mitchell (Minneapolis: University of Minnesota Press, 2000), 1–34.

Mokyr, J. *The Gift of Athena: Historical Origins of the Knowledge Economy* (Princeton: Princeton University Press, 2001).

Moldrich, D. *Bitter Berry Bondage* (Colombo: Ceylon Printers, 1989).

Moore, W. 'The Constitutional Requirements for Tropical Climates, with Special Reference to Temperaments', *Transactions of the Epidemiological Society* 4 (1884–85): 37–8.

Mort, F. *Dangerous Sexualities: Medico-Moral Politics in England Since 1830* (London: Routledge and Kegan Paul, 1987).

Mosse, G. *Nationalism and Sexuality* (New York: Fertig, 1985).

Munasinghe, I. 'The Colombo-Kandy Railway', *The Ceylon Historical Journal* 25 (1978): 239–49.

Nally, D. '"Eternity's Commissioner"; Thomas Carlyle, the Great Irish Famine and the Geopolitics of Travel', *Journal of Historical Geography* 32 (2006): 313–335.

Northrup, D. *Indentured Labour in the Age of Imperialism 1834–1922* (Cambridge: Cambridge University Press, 1995).

Ortner, S. 'Resistence and the Problem of Ethnographic Refusal', *Comparative Studies in Society and History* 37 (1995):173–193.

Osborne T. and Rose, N. 'Governing Cities: Notes on the Spatialization of Virtue' *Environment and Planning D: Society and Space* 17 (1999): 737–60.

O'Tuathail, G. *Critical Geopolitics: The Politics of Writing Global Space* (Minneapolis: University of Minnesota Press, 1996).

Owen, T.C. *First Year's Work on a Coffee Plantation: Oonoogalla Estate, Madulkelle (being the essay which received the second prize from the Ceylon Planters' Association in 1877).* (Colombo. 1877).

Parry, B. *Delusions and Discoveries* (London:Verso, 1998).

Pasteur, L. 'Germ Theory and its Application to Medicine', *Comptes Rendus de l'Académie des Sciences* 86 (1878).

Pasteur, L. 'Extensions of Germ Theory', *Comptes Rendus de l'Académie des Sciences* 88 (1880).

Peiris, G.H. *Development and Change in Sri Lanka: Geographical Perspectives* (New Delhi: Macmillan, 1996).

Perera, N. *Society and Space: Colonialism, Nationalism and Postcolonial Identity in Sri Lanka* (Boulder: Westview, 1998).

Perera, N. *Decolonizing Ceylon: Colonialism, Nationalism, and the Politics of Space in Sri Lanka* (Oxford: Oxford University Press, 1999).

Phillips, R. *Mapping Men and Empire: A Geography of Adventure* (London: Routledge, 1997).

Pippet, G.K. *A History of the Ceylon Police. Vol. 1 1795–1870* (Colombo: Times of Ceylon,1938).

Pieris, R. 'Society and Ideology in Ceylon During a Time of Troubles, 1795–1850, Part 2', *University of Ceylon Review* 9 (1951): 266–79.

Pieris, R. 'Society and Ideology in Ceylon During a "Time of Troubles", 1796–1850, Part 3', *University of Ceylon Review* 10 (1952): 79–102.

Poovey, M. *Making a Social Body: British Cultural Formation, 1830–1864* (Chicago: University of Chicago Press, 1995).

Porter, D. *Health, Civilization and the State* (London: Routledge, 1999).

Porter, R *The Greatest Benefit to Mankind: A Medical History of Humanity from Antiquity to the Present* (London: Harper Collins, 1997).

Prakash, G. *Another Reason: Science and the Imagination of Modern India* (Princeton: Princeton University Press, 1999).

Prakash, G. 'Body Politic in Colonial India', in *Questions of Modernity*, edited by T. Mitchell, 189–222. (Minneapolis: University of Minnesota Press, 2000).

Pratt, M.L. *Imperial Eyes: Travel Writing and Transculturation* (London: Routledge, 1992).

Proceedings of the Ceylon Agricultural Society (Kandy).

Proceedings of the Planters' Association (Kandy)*(PPA)*.

Rabinow, P. *French Modern: Norms and Forms of the Social Environment* (Chicago: University of Chicago Press, 1989).

Ramasamy, P. 'Labour Control and Labour Resistance in the Plantations of Colonial Malaya', in *Plantations, Proletarians and Peasants in Colonial Asia*, edited by E.V. Daniel, H. Bernstein, and T. Brass (London: Frank Cass, 1992), 91–111.

Redfield, P. 'Foucault in the Tropics: Displacing the Panopticon', in *Anthropologies of Modernity: Foucault, Governmentality and Life Politics*, edited by J.X. Inda (Oxford: Blackwell, 2005) 50–79.

Robb, P. (ed.) *The Concept of Race in South Asia* (Delhi: Oxford University Press, 1995).

Roberts, M. 'Indian Estate Labour in Ceylon During the Coffee Period 1830 to 1880', *The Indian Economic and Social History Review* 3 (1966): 1–52, 101–36.

Roberts, M. 'The Master Servant Laws of 1841 and the 1860s and Immigrant Labour in Ceylon', *Ceylon Journal of Historical and Social Studies* 8 (1965): 24–37.

Roberts, M. 'The Impact of the Waste Lands Legislation and the Growth of Plantations on the Techniques of Paddy Cultivation in British Ceylon: A Critique', *Modern Ceylon Studies* 1 (1970): 157–98.

Roberts, M. 'Some Comments on Ameer Ali's Paper', *Ceylon Studies Seminar* 3b (1970/72).

Roberts, M. 'Grain Taxes in British Ceylon, 1832–1878: Theories, Prejudices and Controversies', *Modern Ceylon Studies* 1 (1970): 115–46.

Roberts, M., Rheem, P. Colin-Thome *People Inbetween: The Burghers and the Middle Class in the Transformation of Sri Lanka, 1790–1960* (Ratmalana: Sarvodaya Book Publishing, 1989).

Roberts, M. and Wickremeratne, L.A. 'Export Agriculture in the Nineteenth Century', in *University of Ceylon, History of Ceylon, Volume 3*, edited by K.M. de Silva (Colombo: University of Ceylon, 1973), 101–132.

Rogers, J.D. *Crime, Justice and Society in Colonial Sri Lanka* (London: Curzon Press, 1987).

Rose, N. *Inventing Our Selves: Psychology, Power and Personhood* (Cambridge: Cambridge University Press, 1998).

Rose, N. and C. Novas, 'Biological Citizenship', in *Global Assemblages: Technology, Politics, and Ethics as Anthropological Problems*, edited by A. Ong and S. Collier (Oxford: Blackwell, 2005), 439–463.

Rose-Redwood, R. 'Governmentality, Geography and the Geo-Coded World', *Progress in Human Geography* 30 (2006): 469–86.

Rosenberg, C.E. *Healing and History* (New York: Science History Publications, 1979).

Rutherford, J. *Forever England: Reflections on Masculinity and Empire* (London: Lawrence and Wishart, 1997).

Rutherford, P. 'The Entry of Life into History', in *Discourses of the Environment*, edited by E. Darier (Oxford: Blackwell, 1998), 37–61.

Sabonadiere, W. *The Coffee Planter of Ceylon* (Guernsey: MacKenzie, Son and Le Patourel, 1866).

Said, E. *Orientalism* (New York: Vintage, 1979).

Samaraweera, V. 'British Justice and the "Oriental Peasantry": The Working of the Colonial Legal System in Nineteenth Century Sri Lanka', in *British Imperial Policy in India and Sri Lanka, 1858–1912*, edited by R.I. Crane and G. Barrier (Columbia: South Asia Books, 1981), 107–41.

Scheper-Hughes, N. *Death Without Weeping: The Violence of Everyday Life in Brazil* (Berkeley: University of California Press, 1992).

Schivelbusch, W. *Tastes of Paradise* (New York: Vintage, 1993).

Scott, D. 'Colonial Governmentality', in *Anthropologies of Modernity: Foucault, Governmentality and Life Politics* edited by J. Inda (Oxford: Blackwell, 2005), 23–49.

Scott, J. *Weapons of the Weak: Everyday Forms of Peasant Resistance* (New Haven: Yale University Press, 1985).

Scott, J. *Seeing Like a State: How Certain Schemes to Improve the Human Condition Have Failed* (New Haven: Yale University Press, 1998).

Sessional Papers (Papers Laid Before the Legislative Council of Ceylon).

Sirr, H.C. *Ceylon and the Cingalese. Vol. 2* (London: William Sholbert, 1850).

Sri Lanka National Archive (SLNA) Despatches.

Smith, S. 'Sugar's Poor Relation: Coffee Planting in the British West Indies, 1720–1833', *Slavery and Abolition* 19 (1998): 151–72.

Snodgrass, D. *Ceylon: An Export Economy in Transition* (Homewood, Illinois: Richard Irwin, 1966).

Somasekara, T. et al. (eds), *Arjuna's Atlas of Sri Lanka* (Dehiwala: Arjuna's Consulting. 1997).

Spurr, D. *The Rhetoric of Empire: Colonial Discourse in Journalism, Travel Writing and Imperial Administration* (Durham: Duke University, 1993).

Stedman Jones, G. *Outcast London: A Study in the Relationship between Classes in Victorian Society* (Oxford: Oxford University Press, 1971).

Stepan, N. *The Idea of Race in Science, Great Britain, 1800–1960* (London: Macmillan, 1982).

Stepan, N.L. 'Race and Gender: the Role of Analogy in Science', in *Anatomy of Racism*, edited by D. Goldberg (Minneapolis: University of Minnesota, 1990), 38–57.

Stepan, N.L. *Picturing Tropical Nature* (London: Reaktion, 2001).

Stewart, L. 'Louisiana Subjects: Power, Space and the Slave Body', *Ecumene* 2 (1995): 227–45.

Stoler, A.L. *Carnal Knowledge and Imperial Power: Race and the Intimate in Colonial Rule* (Berkeley: University of California Press, 2002).

Stoler, A.L. 'Plantation Politics and Protest on Sumatra's East Coast', *Journal of Peasant Studies* 13 (1986): 212–33.

Stoler, L.A. 'Rethinking Colonial Categories: European Communities and the Boundaries of Rule', in *Colonialism and Culture*, edited by N.B. Dirks, 119–52. (Ann Arbor: University of Michigan, 1991).

Stoler, A.L. and F. Cooper, 'Between Metropole and Colony: Rethinking a Research Agenda', in *Tensions of Empire: Colonial Cultures in a Bourgeois World*, edited by F. Cooper and A.L. Stoler (Berkeley: University of California Press, 1997), 1–58.

Stoler, A.L. 'Sexual Affronts and Racial Frontiers: European Identities and the Cultural Politics of Exclusion in Colonial Southeast Asia', in *Tensions of Empire: Colonial Cultures in a Bourgeois World*, edited by F. Cooper and A.L. Stoler (Berkeley: University of California Press, 1997), 198–237.

Stoler, L.A. *Race and the Education of Desire: Foucault's History of Sexuality and the Colonial Order of Things* (Durham: Duke University Press, 1995).

Tausig, M. *The Devil and Commodity Fetishism in South America* (Chapel Hill: 1980).

Taylor, J. *Papers of James Taylor (1835–92).* (MSS. 15908–10). National Library of Scotland. Department of Manuscripts, 1851–92.

Thomas, N. *Colonialism's Culture: Anthropology, Travel, and Government* (Cambridge. Polity, 1994).

Thomas, N. and Eves, R. *Bad Colonists: The South Seas Letters of Vernon Lee Walker and Louis Becke* (Durham: Duke University Press, 1999).

Thorn, J. *Le Café* (Koln: Taschen, 2001).

Thorne, S. '"The Conversion of Englishmen and the Conversion of the World Inseparable": Missionary Imperialism and the Language of Class in Early Industrial Britain', in *Tensions of Empire: Colonial Cultures in a Bourgeois World*, edited by F. Cooper and A.L. Stoler (Berkeley: University of California Press, 1997), 238–62.

Thrift, N. *Knowing Capitalism* (London: Sage, 2005).

Times of Ceylon.

Tinker, H. *A New System of Slavery: The Export of Indian Labour Overseas 1830–1920* (Oxford: Oxford University Press, 1974).

Topik, S. 'The Integration of the World Coffee Market', in *The Global Coffee Economy in Africa, Asia and Latin America, 1500–1989*, edited by W.G. Clarence-Smith and S. Topik (Cambridge: Cambridge University Press, 2003), 21–49.

Topik, S. and Clarence-Smith W.G. 'Introduction: Coffee and Global Development', in *The Global Coffee Economy in Africa, Asia and Latin America, 1500–1989*, edited by W.G. Clarence-Smith and S. Topik (Cambridge: Cambridge University Press, 2003), 1–20.

Tosh, J. 'Domesticity and manliness', in *Manful Assertions: Masculinities in Britain since 1800*, edited by M. Roper and J. Tosh (London: Routledge. 1991) 44–73.

Tosh, J. *Manliness and Masculinities in Nineteenth-Century Britain: Essays on Gender, Family and Empire* (London: Pearson Longman, 2005).

Tuchscherer, M. 'Coffee in the Red Sea Area From the Sixteenth to the Nineteenth Century', in *The Global Coffee Economy in Africa, Asia and Latin America, 1500–1989*, edited by W.G. Clarence-Smith and S. Topik (Cambridge: Cambridge University Press, 2003), 50–66.

Twining, W. *Clinical Illustrations of the More Important Diseases of Bengal, with the Result of an Inquiry into Their Pathology and Treatment* (Calcutta: Baptist Mission, 1832).

Uragoda, C.G. *A History of Medicine in Sri Lanka* (Colombo: Middleway Limited, 1987).

Van Dort, W.G. 1869–70. 'Gampola Hospital Report 1869', reprinted in *PPA*, 1869–70, 54–55.

Vanden Driesen, I.H. 'Coffee Cultivation in Ceylon (1)', *The Ceylon Historical Journal* 3 (1953): 31–61.

Vanden Driesen, I.H. 'Coffee Cultivation in Ceylon (2)', *The Ceylon Historical Journal* 3 (1953): 156–72.

Vanden Driesen, I.H. 'Some Trends in the Economic History of Ceylon in the "Modern" Period', *Ceylon Journal of Historical and Social Studies* 3 (1960): 1–17.

Vanden Driesen, I.H. 'Land Sales Policy and Some Aspects of the Problem of Tenure, 1836–86. Part 2', *University of Ceylon Review* 15 (1957): 36–52.

Vanden Driesen, I.H. *Indian Plantation Labour in Sri Lanka: Aspects of the History of Immigration in the 19th Century* (Nedlands: University of Western Australia, 1982).

Ward, M. 'Research on the Life History of *Hemileia Vastatrix*, the Fungus of the "Coffee-Leaf" Disease', *Journal of the Linnean Society of London – Botany* 19 (1882).

Watts, M. 'Enclosure: A Modern Spatiality of Nature', in *Envisioning Human Geography*, edited by P. Cloke, P. Crang, and M. Goodwin (London: Arnold, 2004), 8–64.

Watts, S. *Epidemics and History: Disease, Power and Imperialism* (New Haven: Yale University Press, 1997).

Weatherstone, J. *The Pioneers: The Early British Tea and Coffee Planters and Their Way of Life, 1825–1900* (London: Quiller Press, 1986).

Webb, J.L.A. *Tropical Pioneers: Human Agency and Ecological Change in the Highlands of Sri Lanka, 1800–1900* (Athens: Ohio University Press, 2002).

Wenzlhuemer, R. 'The Sinhalese Contribution to Estate Labour in Ceylon, 1881–1891', *Journal of the Economic and Social History of the Orient* 48 (2005): 442–58.

Wesumperuma, D. *Indian Immigrant Plantation Workers in Sri Lanka: A Historical Perspective 1880–1910* (Kelaniya: Vidyalankara Press, 1886).

Wesumperuma, D. 'The Evictions Under the Paddy Tax and Their Impact on the Peasantry of Walapane, 1882–1885', *The Ceylon Journal of Historical and Social Studies* 10 (1970): 131–48.

Wickremeratne, L.A. 'Grain Consumption and Famine Conditions in Late Nineteenth Century Ceylon', *The Ceylon Journal of Historical and Social Studies* NS 3 (1973): 28–53.

Wickramasinghe, N. *Dressing the Colonised Body: Politics, Clothing and Identity in Colonial Sri Lanka* (New Delhi: Orient Longman, 2003).

Wiener, M.J. *Reconstructing the Criminal: Culture, Law and Policy in England, 1830–1914* (Cambridge: Cambridge University Press, 1990).

Williams, R. *Culture and Society, 1780–1950* (Harmondsworth: Penguin, 1963).

Wilson, K. *The Island Race: Englishness, Empire and Gender in the Eighteenth Century* (London: Routledge, 2003).

Wolf, E. 'Specific Aspects of Plantation Systems in the New World', in *Plantation Systems of the New World*, edited by A. Palerm and V. Rubin (Washington: Pan American Union, 1959), 136–46.

Worboys, M. 'Germs, Malaria and the Invention of Mansonian Tropical Medicine: From "Diseases in the Tropics" to "Tropical Diseases"', in *Warm Climates and Western Medicine*, edited by D. Arnold (Amsterdam: Rodopi, 1996), 181–207.

Young, I.M. *Justice and the Politics of Difference* (Princeton: Princeton University Press, 1990).

Young, R. *Colonial Desire: Hybridity in Theory, Culture and Race* (London: Routledge, 1995).

Index

211